Springer Theses

Recognizing Outstanding Ph.D. Research

Aims and Scope

The series "Springer Theses" brings together a selection of the very best Ph.D. theses from around the world and across the physical sciences. Nominated and endorsed by two recognized specialists, each published volume has been selected for its scientific excellence and the high impact of its contents for the pertinent field of research. For greater accessibility to non-specialists, the published versions include an extended introduction, as well as a foreword by the student's supervisor explaining the special relevance of the work for the field. As a whole, the series will provide a valuable resource both for newcomers to the research fields described, and for other scientists seeking detailed background information on special questions. Finally, it provides an accredited documentation of the valuable contributions made by today's younger generation of scientists.

Theses are accepted into the series by invited nomination only and must fulfill all of the following criteria

- They must be written in good English.
- The topic should fall within the confines of Chemistry, Physics, Earth Sciences, Engineering and related interdisciplinary fields such as Materials, Nanoscience, Chemical Engineering, Complex Systems and Biophysics.
- The work reported in the thesis must represent a significant scientific advance.
- If the thesis includes previously published material, permission to reproduce this must be gained from the respective copyright holder.
- They must have been examined and passed during the 12 months prior to nomination.
- Each thesis should include a foreword by the supervisor outlining the significance of its content.
- The theses should have a clearly defined structure including an introduction accessible to scientists not expert in that particular field.

More information about this series at http://www.springer.com/series/8790

Yutaro Iiyama

Search for Supersymmetry in pp Collisions at $\sqrt{s} = 8$ TeV with a Photon, Lepton, and Missing Transverse Energy

Doctoral Thesis accepted by Carnegie Mellon University, New York, USA

 Springer

Yutaro Iiyama
Laboratory for Nuclear Science
Massachusetts Institute of Technology
Cambridge, Massachusetts, USA

ISSN 2190-5053   ISSN 2190-5061   (electronic)
Springer Theses
ISBN 978-3-319-86448-8   ISBN 978-3-319-58661-8   (eBook)
DOI 10.1007/978-3-319-58661-8

Printed on acid-free paper

This Springer imprint is published by Springer Nature
The registered company is Springer International Publishing AG
The registered company address is: Gewerbestrasse 11, 6330 Cham, Switzerland

Supervisor's Foreword

The standard model of particle physics is an enormously successful theory describing the interactions of all known elementary particles: quarks, leptons, and gauge bosons. Developed over the past 50 years, starting with the quark model in the 1960s, the discovery of the charm quark in 1974, the τ lepton in experiments from 1974 to 1977, the bottom quark in 1977, the W and Z bosons in 1983, the top quark in 1995, and culminating in the discovery of the Higgs boson in 2012, there is to date no experimental evidence contradicting the predictions of the standard model. Although it is successful in describing all phenomena at the subatomic scale, it is not a complete "theory of everything" that can explain all known observations. For example, no particle exists in the standard model that constitutes a possible candidate for dark matter, which makes up about one quarter of the energy-mass content of the universe. The quest for understanding the nature of dark matter is one reason why physicists at the CERN Large Hadron Collider (LHC) are searching for yet-unknown particles, which can pave the way to postulate theories beyond the standard model.

Among the various candidates of such theories, the extension of the standard model with supersymmetry (SUSY) is considered one of the most promising possibilities. SUSY is a hypothetical symmetry that predicts a partner particle for each of the standard model particles, thus doubling the number of known elementary particles. The newly introduced particles share the same properties, such as electric charge, with their standard model counterparts, but differ in their spin – the particle's intrinsic angular momentum – and are typically much heavier than their corresponding standard model partners. These SUSY particles with large masses can be produced in high-energy particle collisions such as the proton-proton collisions taking place at the LHC. In most cases, the produced particles would immediately decay into known combinations of standard model particles plus at least one dark matter particle. The latter does not interact with regular matter, and escapes undetected a particle detector, such as the Compact Muon Solenoid (CMS) experiment at the LHC. Thus, a standard method to search for the evidence of SUSY particles produced in the LHC collisions is to look for specific combinations of standard model particles with sufficient energies, accompanied by a large imbalance in the measured momenta of the outgoing particles, which would be caused by the dark matter particle escaping detection.

A search for SUSY utilizing a final state consisting of a photon, an electron, or a muon and a large momentum imbalance has been performed in this Ph.D. thesis by Dr. Yutaro Iiyama under my supervision in the Department of Physics at Carnegie Mellon University. This search probes a class of SUSY models that has not yet been studied extensively. In contrast to many other SUSY searches performed so far, this choice of final state allows the search to be sensitive to an important SUSY production mode that does not involve the SUSY partners of quarks and gluons (squarks and gluinos), but the supersymmetric partners of the heavy W and Z bosons mediating the weak force. SUSY searches involving the strong production of squarks and gluinos at the LHC are expected to be more frequent than other SUSY production modes and were thus targeted in the first SUSY searches by the LHC experiments. Since no evidence of SUSY particles was found through the strong production mechanism, searches for the more rare, electroweak production via the SUSY partners of the W and Z bosons are attempted.

This Ph.D. thesis is one of the first searches contributing to the class of "electroweakino" SUSY searches at the LHC utilizing pp collision data taken at a center-of-mass energy of $\sqrt{s} = 8\,\text{TeV}$ in 2012 by the CMS experiment. Unfortunately, no evidence of SUSY production was found in this thesis and the result puts constraints on the possible parameter space of the class of studied SUSY models. In particular, a lower limit on the possible mass of the SUSY partners of the electroweak W and Z gauge bosons is drawn regardless of the masses of the squarks and gluinos.

This thesis entitled "Search for Supersymmetry in pp Collisions at $\sqrt{s} = 8\,\text{TeV}$ with a Photon, Lepton, and Missing Transverse Energy" is organized in seven chapters as follows:

Chapter 1 provides a short introduction into particle physics and the LHC accelerator, leading to the motivation for this search for SUSY with a photon and lepton in the final state plus missing energy or momentum.

Chapter 2 describes the standard model and gives a very extensive introduction into the topic of supersymmetry, which goes beyond the scope of describing SUSY for an experimental thesis. I would consider this discussion of SUSY phenomenology adequate as part of a thesis by a theory student who worked on a theoretical study of supersymmetry phenomenology.

Chapter 3 is a nice discussion of the LHC and the main elements of the CMS detector including the trigger system and the reconstruction of photon, electron, muon, and jet objects.

This is followed by a discussion of the data collection and event selection in Chap. 4, which also contains a description of the simulation of signal and background Monte Carlo samples.

Chapter 5 is the main element of the thesis, describing in detail the data analysis involved in searching for SUSY and the study of backgrounds faking the signal.

As in every search for physics beyond the standard model, the goal is to suppress standard model processes yielding the same final state and predict the remaining standard model backgrounds as precisely as possible. For some of the backgrounds, such as jets or electrons faking photons, established techniques exist within CMS, and Dr. Iiyama performed these measurements in a way that they were also useful for other SUSY analyses in CMS. To explain the excellence of this thesis, let me briefly mention that one of the main backgrounds at large missing energy is the $W\gamma$ production, which yields the same final state signature as the SUSY signal being searched for. Using the Monte Carlo expectation for this background would not have allowed for a reliable background prediction. To tackle this main background of the analysis, Dr. Iiyama came up with a novel method to predict the background from data using a two-component fit with a kinematic variable, $\Delta\phi(\ell, E_T^{\mathrm{miss}})$, that also allowed him to better pin down the fake lepton background prediction—a very clever idea for a robust method to deal with this major background issue.

Chapter 6 discusses the results of this analysis and the limit-setting procedure, including the interpretation of this search in the framework of SUSY signal models.

Finally, Chap. 7 offers concluding remarks and an outlook to future SUSY searches utilizing the photon plus lepton final state.

With the current run of the LHC at the higher center-of-mass energy of $\sqrt{s} = 13\,\mathrm{TeV}$, this photon plus lepton analysis plays an important role in the next round of searches for electroweak production of SUSY particles at the LHC with one of my current graduate students performing this analysis.

Professor of Physics   Manfred Paulini
October 2016

Acknowledgments

The work described in this thesis owes almost its entirety to the dedicated effort of the Large Hadron Collider accelerator team and the Compact Muon Solenoid collaboration. I would like to acknowledge the excellent performance of the machine and the detector throughout Run I and congratulate all who were involved from the conception of the LHC program for their hard work culminating in this world's most exciting scientific apparatus. I would also like to express my gratitude to the CMS Electromagnetic Calorimeter group and particularly to its leadership, Tommaso Tabarelli De Fatis, Dave Barney, and David Petyt, for guiding me into the world of collider physics collaboration. My main roles in the ECAL group, as an online operator and the DQM developer, were hugely assisted by André David and Cristina Biino, who never failed to amaze me with their competence and the deep insight to the inner workings of the detector, and Emanuele Di Marco and Giuseppe Della Ricca, whose extraordinary professionalism is imprinted in the DQM software and infrastructure.

My analysis work was performed within the SUSY Photon subgroup, where the relative smallness of the group helps lively discussions. The leaders while I was involved in the analysis, Manfred Paulini, Dave Mason, Christian Autermann, Yuri Gerstein, and Yurii Maravin, guided the group fantastically, even during the time-constrained conference preparations. Some of the ideas in this thesis grew out of the discussions in the subgroup, in particular with Andrew Askew, Dave Morse, Brian Francis, and Rachael Yohay. The SUSY group at large has also been very helpful in pushing the analysis forward. I would like to thank particularly the supportive leadership of Keith Ulmer and Filip Moortgat. The CMS-internal analysis review committee members, Artur Apresyan, Isabell Melzer-Pellman, and Hartmut Stadie, gave numerous invaluable comments and suggestions, some of which are directly implemented in this thesis. The motivation of the analysis and its interpretation received important contributions from David Shih.

Looking more broadly to my training and education in the graduate school as a whole, I simply cannot express how grateful I am to my advisor Manfred Paulini. Manfred has always let me think and decide on the analysis and other activities that we worked on together, and would always give me a second chance at many occasions that I made wrong decisions. It has been such a pleasure to be his student, and I know who I want to be like if I end up in a position to guide a student. I would also like to thank other Carnegie Mellon group members Virginia Azzolini, Aristotle Calamba, Ben Carlson, and Menglei Sun for being wonderful coworkers.

During my graduate studies, I had the privilege of staying at CERN for 3 years and 8 months. Being where the action is was a great experience on its own, but I would not have felt so excited every day in my life at Geneva were it not for my fabulous friends. The circle has greatly expanded from a group of ECAL chalet students, so I will refrain from listing out every one of the "coolest kids at CERN," but would like to thank particularly Andrew Brinkerhoff, Keisuke Yoshihara, Nil Valls, Nathan Kellams, and Michael Planer, who were my roommates at various times.

The continuous encouragement and support from my family has always brought me a sense of security. I probably felt relatively at ease being in a foreign country because of my past experience doing so, also thanks to my family, but there are times when knowing that there is somewhere to come back to makes a big difference. Finally, my deepest appreciation is to Sachi Inukai, for everything she does brightens my day.

Contents

Chapter 1
Introduction

Particle physics is an endeavor to explain the natural world from a single principle. It is often said that the discipline takes root in the classic question: "What are we made of?"

The standard model of particle physics in its current form, which has been known for more than 40 years, already answers this question to some extent. According to the model, "we" are made of matter particles, which are the electrons, up quarks, and down quarks that interact via the photon and gluon fields. The quarks form the nucleons, which then bind together to form nuclei. Nuclei and the electrons form atoms, from which molecules are made. The mass of the electron is due to its interaction with the Higgs field. The quarks are also made massive by the Higgs field, but the nucleons owe most of its mass to the self-interaction energy of the quarks and gluons. The standard model has been subjected to numerous experimental tests [1], and each time it has proven to be consistent with observation. The most notable of its confirmations in the recent years is the discovery of the Higgs boson [2, 3] in 2012.

Despite the remarkable success of the standard model, the fundamental structure of nature is far from explained. Why are the constituents of the standard model what they are? How are the parameters of the model determined? And perhaps most pressingly, what is out there, i.e., what is the dark matter that constitutes about a quarter of the mass energy of the universe? There is mounting evidence of the existence of dark matter from astronomy, yet there is none so far in the realm of particle physics. A clue to these problems could come from an observation of an unexplained phenomenon, yet the results of experiment after experiment keep confirming the standard model.

While lacking concrete guidance on how particle physics should proceed, the prevailing assumption is that the standard model is an effective theory valid for the length and energy scales that have been probed so far, and a more fundamental description of nature exists for a shorter-length and higher-energy scale. Such a hierarchy of scales is seen everywhere in nature; the details of molecular interactions

© Springer International Publishing AG 2017
Y. Iiyama, *Search for Supersymmetry in pp Collisions at* $\sqrt{s} = 8$ TeV *with a Photon, Lepton, and Missing Transverse Energy*, Springer Theses,
DOI 10.1007/978-3-319-58661-8_1

are not necessary to describe the general properties of a fluid, and in chemistry, which is the theory of molecular interactions, the nucleons don't play an active role.

The supersymmetric extension of the standard model is one candidate of such a higher-scale theory. In this extension, particles of the standard model are elevated to multiplets of particles sharing the same quantum numbers but with spins differing by a half-unit. The supersymmetric standard model does not only offer a natural candidate for a dark matter particle, but it also solves another important problem of the standard model, namely the naturalness of the Higgs boson mass. The naturalness problem is a conflict of the observed Higgs boson mass, at about 126 GeV, with the fact that there is no mechanism in the standard model to stop the Higgs field from acquiring mass at much higher-energy scales. The supersymmetric standard model is equipped with such a mechanism quite independently from the details of the specific model. There are many more virtues of supersymmetry aside from those mentioned, making it one of the most favored scenarios for physics beyond the standard model.

A separation of energy scales often implies the existence of a massive particle. At the lower-energy scale, the strength of the interaction involving such a particle is suppressed by powers of the typical momentum exchanged in the interaction divided by its mass, thus hiding the particle and, with it, the higher-energy scale from the lower-scale description of nature. If, then, it is possible to produce and study particles of an energy scale higher than the electroweak scale, i.e., the scale of the standard model, particle physics can break through into new territory.

The CERN Large Hadron Collider (LHC) was indeed built with the goal of producing such unknown particles. With the capability of delivering proton–proton collisions at a center-of-mass energy of 14 TeV, it is the highest-energy particle collider to date. The high-energy, high-intensity proton–proton interactions at the LHC open a window to the phenomena beyond the electroweak scale. A general-purpose particle detector, the Compact Muon Solenoid (CMS), is located at one of the LHC collision points to study the results of the proton–proton hard scattering. Its excellent energy and momentum resolution in observing particles emerging from the collision point enables an almost complete reconstruction of the scattering. By recording the collision events and studying the outgoing particles, or the final states, through statistical analyses, it is possible to infer the production of unknown heavy particles, which can signify the correctness of the supersymmetric standard model or some other theory of new physics.

This thesis presents a search for supersymmetry in a final state including a photon, an electron or a muon, and an undetected high-energy particle. This final-state signature is simple and robust, and yet is highly discriminating against typical signals from proton–proton scattering emerging from interactions of the standard model. Therefore, this search is able to probe a region of the parameter space for a broad class of supersymmetric standard models that are otherwise not reachable.

The organization of this thesis is as follows. Chapter 2 gives an overview of the theory of the standard model and its supersymmetric extension, with an emphasis on models based on gauge-mediated supersymmetry breaking. A brief survey of the current status of the searches for supersymmetry is given at the end of the chapter.

The experimental apparatuses, i.e., the LHC and CMS, are described in Chap. 3, while Chap. 4 details the collection of the data. Chapter 5 discusses the analysis methods. The results of the search and its interpretations are presented in Chap. 6. Finally, the thesis is concluded in Chap. 7.

References

1. ALEPH, CDF, D0, DELPHI, L3, OPAL, SLD Collaborations, the LEP Electroweak Working Group, the Tevatron Electroweak Working Group, and the SLD Electroweak and Heavy Flavour Groups: Precision electroweak measurements and constraints on the standard model (2010). http://lepewwg.web.cern.ch/LEPEWWG
2. CMS Collaboration: Observation of a new boson at a mass of 125 GeV with the CMS experiment at the LHC. Phys. Lett. B **716**, 30–61 (2012). doi:10.1016/j.physletb.2012.08.021. arXiv: 1207.7235 [hep-ex], CMS-HIG-12-028, CERN-PH-EP-2012-220
3. ATLAS Collaboration: Observation of a new particle in the search for the Standard Model Higgs boson with the ATLAS detector at the LHC. Phys. Lett. B **716**, 1–29 (2012). doi:10.1016/j.physletb.2012.08.020. arXiv: 1207.7214 [hep-ex], CERN-PH-EP-2012-218

Chapter 2
The Standard Model and Its Supersymmetric Extension

2.1 The Lagrangian of the Standard Model

The standard model of particle physics (SM) is a Lorentz-invariant quantum field theory (QFT) that is highly successful in describing the properties of subatomic particles and their interactions. The Lagrangian of the theory can be split into four distinct sectors:

$$\mathcal{L}_{\mathrm{SM}} = \mathcal{L}_{\mathrm{kin.}} + \mathcal{L}_{\mathrm{Y}} + \mathcal{L}_{\mathrm{YM}} + \mathcal{L}_{\mathrm{V}}. \tag{2.1}$$

The first term of Eq. (2.1) describes propagating fields, with

$$\mathcal{L}_{\mathrm{kin.}} = iq^{\dagger}\overline{\sigma}^{\mu}\nabla_{\mu}q + i\bar{u}^{\dagger}\overline{\sigma}^{\mu}\nabla_{\mu}\bar{u} + i\bar{d}^{\dagger}\overline{\sigma}^{\mu}\nabla_{\mu}\bar{d} + i\ell^{\dagger}\overline{\sigma}^{\mu}\nabla_{\mu}\ell + i\bar{e}^{\dagger}\overline{\sigma}^{\mu}\nabla_{\mu}\bar{e} + (\nabla_{\mu}h)^{*}(\nabla^{\mu}h), \tag{2.2}$$

where fermions q, $\bar{u}$, $\bar{d}$, ℓ, and $\bar{e}$ are all two-component left-chiral spinor fields, and the Higgs field h is a scalar. The bar in the field names is conventional and does not imply any mathematical operation. The set of 2×2 matrices $\overline{\sigma}^{\mu} = (I_2, -\sigma^1, -\sigma^2, -\sigma^3)$, where σ^i are Pauli matrices, combines left- and right-chiral representations of the Lorentz group to form a vector representation. In this expression and all that follows, summation over paired indices is implied unless otherwise stated. Spinors are contracted following the convention, e.g., $q^{\dagger}\overline{\sigma}^{\mu}\nabla_{\mu}q :=$ $(q^{\dagger})_{\dot{\alpha}}(\overline{\sigma}^{\mu})^{\dot{\alpha}\beta}(\nabla_{\mu}q)_{\beta}$.

A gauge transformation with gauge group $G_{\mathrm{SM}} = \mathrm{U}(1) \times \mathrm{SU}(2) \times \mathrm{SU}(3)$ acts on the fields in Eq. (2.2). The fields can be categorized by the representation of G_{SM} they belong to, as given in Table 2.1. Representations are denoted by 3-tuples. The first element of the 3-tuple is either **3**, $\overline{\mathbf{3}}$, or **1** and indicates, respectively, the fundamental representation, its conjugate, and the singlet (trivial) representation of SU(3). The second element is for SU(2) and is either **2** or **1**, corresponding to the fundamental and the singlet representations. The last element is the hypercharge, or

© Springer International Publishing AG 2017
Y. Iiyama, *Search for Supersymmetry in pp Collisions at* $\sqrt{s} = 8$ TeV *with a Photon, Lepton, and Missing Transverse Energy*, Springer Theses, DOI 10.1007/978-3-319-58661-8_2

Table 2.1 Fermion and Higgs fields of the standard model

Field	Name	G_{SM} representation
q	Quark doublet	$(\mathbf{3}, \mathbf{2}, \frac{1}{6})$
$\bar{u}$	Up singlet	$(\bar{\mathbf{3}}, \mathbf{1}, -\frac{2}{3})$
$\bar{d}$	Down singlet	$(\bar{\mathbf{3}}, \mathbf{1}, \frac{1}{3})$
ℓ	Lepton doublet	$(\mathbf{1}, \mathbf{2}, -\frac{1}{2})$
$\bar{e}$	Lepton singlet	$(\mathbf{1}, \mathbf{1}, 1)$
h	Higgs	$(\mathbf{1}, \mathbf{2}, \frac{1}{2})$

See text for the notation

the scale factor on the phase that each field gains under U(1) transformation. Note that the hypercharge assignments differ from what are used in some literatures by a factor 2.

From Table 2.1 it is evident that only q is in nontrivial representations of all three component groups of G_{SM}. The covariant derivative of q is defined as

$$\nabla_\mu q_{ai} = \partial_\mu q_{ai} + i g_3 G_\mu^A \frac{(\lambda_A)_{ab}}{2} q_{bi} + i g_2 W_\mu^I \frac{(\sigma_I)_{ij}}{2} q_{aj} + i g_1 Y B_\mu q_{ai} \tag{2.3}$$

where $a, b = 1, 2, 3$; $i, j = 1, 2$; $A = 1, \ldots, 8$; and $I = 1, 2, 3$. λ_A and σ_I are Gell-Mann and Pauli matrices normalized so that $\text{Tr}(\lambda_A \lambda_B) = 2\delta_{AB}$, $\text{Tr}(\sigma_I \sigma_J) = 2\delta_{IJ}$. $Y = \frac{1}{6}$ is the hypercharge of q. All other covariant derivatives can be constructed similarly using coupling constants g_1, g_2, and g_3 and vector potentials G_μ^A, W_μ^I, and B_μ. The fact that the vector potentials are common among different covariant derivatives causes the interactions between the fields in the SM. Vector potentials are collectively called gauge bosons and are said to mediate gauge interactions; G_μ^A, the gluon, is responsible for the strong interaction, and W_μ^I and B_μ for the electroweak interaction.

The second term of Eq. (2.1) defines the Yukawa interactions between the fermions and the Higgs field:

$$\mathcal{L}_Y = y^u h \circ q\bar{u} - y^d h^* q\bar{d} - y^e h^* \ell\bar{e} + \text{c.c.} \tag{2.4}$$

The operator $\circ$ is the invariant contraction of two fundamental representations of SU(2): $h \circ q = h_1 q_2 - h_2 q_1$. Given the specific combinations of the fields, it is useful to assign additive quantum numbers B and L, respectively, called baryon and lepton numbers, to the fermion fields. By convention, $(B, L) = (1/3, 0)$ for q, $(-1/3, 0)$ for $\bar{u}$ and $\bar{d}$, $(0, 1)$ for ℓ and $(0, -1)$ for $\bar{e}$, which makes the sum of B and L vanish for each term in Eq. (2.4). The sign-inverted values are assigned to the hermitian conjugates.

The fermion fields actually come in three "generations"; for each of q, $\bar{u}$, $\bar{d}$, ℓ, and $\bar{e}$, there are two more fields with identical representation of G_{SM}. Since the covariant derivatives do not mix generations but the Yukawa couplings do, y^u, y^d, and y^e in Eq. (2.4) should be regarded as 3×3 matrices summing over three generations of the two fermions in each term.

The third term of Eq. (2.1) contains field-strength tensors of the gauge bosons,

$$\mathcal{L}_{\text{YM}} = -\frac{1}{4}G^A_{\mu\nu}G^{A\mu\nu} - \frac{1}{4}W^I_{\mu\nu}W^{I\mu\nu} - \frac{1}{4}B_{\mu\nu}B^{\mu\nu}, \tag{2.5}$$

where

$$G^A_{\mu\nu} = \partial_\mu G^A_\nu - \partial_\nu G^A_\mu - g_3 f^{ABC}G^B_\mu G^C_\nu, \tag{2.6}$$

$$W^I_{\mu\nu} = \partial_\mu W^I_\nu - \partial_\nu W^I_\mu - g_2 \epsilon^{IJK}W^J_\mu W^K_\nu, \tag{2.7}$$

$$B^I_{\mu\nu} = \partial_\mu B^I_\nu - \partial_\nu B^I_\mu. \tag{2.8}$$

f^{ABC} is the structure constant of the Gell-Mann matrices, and ϵ^{IJK} is the Levi-Civita tensor.

The last term of Eq. (2.1) is the Higgs potential

$$\mathcal{L}_{\text{V}} = -V(h), \tag{2.9}$$

which will be discussed in detail in Sect. 2.2.

2.2 Electroweak Symmetry Breaking

While the Lagrangian of the SM is invariant under G_{SM} gauge transformation, the observed physical states in the universe do not seem to be SU(2)-symmetric. For instance, neutrinos, which can be regarded as excitations of one component of ℓ, are states distinct from charged leptons, the excitations of the other component. Since observed physical phenomena are described as fluctuations about the vacuum, the implication is that the SU(2) symmetry is spontaneously broken. In fact, it is the larger subgroup SU(2)×U(1) of G_{SM}, electroweak symmetry, that is broken, leaving another U(1) as the electromagnetic gauge symmetry of the vacuum.

The breaking of electroweak symmetry is caused by the vacuum condensation of the Higgs field. The Higgs potential $V(h)$ contains quadratic and quartic terms of h. Denoting these terms as

$$V(h) = \mu^2 h^* h + \lambda (h^* h)^2, \tag{2.10}$$

if $\mu^2 < 0$, the classical vacuum expectation value (VEV) of h is

$$|\langle h \rangle| = \sqrt{\frac{-\mu^2}{2\lambda}}. \tag{2.11}$$

With a choice of the basis of the SU(2) representation such that the VEV of h is along the real axis of the lower component,

$$\langle h \rangle = \frac{1}{\sqrt{2}}\begin{pmatrix} 0 \\ v \end{pmatrix}, \tag{2.12}$$

where $v/\sqrt{2}$ is the VEV in the full quantum theory. The Higgs field h can be expressed in a polar representation with radial and phase degrees of freedom as

$$h = \frac{1}{\sqrt{2}} U(\varphi) \begin{pmatrix} 0 \\ v + s \end{pmatrix} := \frac{1}{\sqrt{2}} e^{i\varphi^I \sigma_I} \begin{pmatrix} 0 \\ v + s \end{pmatrix}, \tag{2.13}$$

where φ and s are now the propagating degrees of freedom about the vacuum. Substituting this expression into the kinetic part of Eq. (2.9) and redefining the weak gauge fields as

$$W'_\mu{}^I \sigma_I := W_\mu^I \sigma_I + \frac{2i}{g_2} U(\varphi) \partial_\mu U^\dagger(\varphi), \tag{2.14}$$

the kinetic term for the Higgs field is

$$(\nabla_\mu h)^* (\nabla^\mu h) = \frac{1}{2} \partial_\mu s \partial^\mu s + \frac{1}{4} (v + s)^2 \left[g_2{}^2 W_\mu^+ W^{-\mu} + \frac{g_1{}^2 + g_2{}^2}{2} Z_\mu Z^\mu \right], \tag{2.15}$$

where

$$W_\mu^\pm := \frac{1}{\sqrt{2}} \left(W'_\mu{}^1 \mp i W'_\mu{}^2 \right) \tag{2.16}$$

and

$$Z_\mu := \cos \theta_W W'_\mu{}^3 - \sin \theta_W B_\mu \tag{2.17}$$

with the weak mixing angle or Weinberg angle θ_W defined by $\tan \theta_W := g_1/g_2$.

To understand the implications on $\mathcal{L}_Y$ of this re-expression of the Higgs Lagrangian, it is instructive to interpret Eq. (2.14) as an actual gauge transformation. From this perspective, each configuration of h defines the so-called unitary gauge, in which weak bosons are manifestly massive and the goldstone bosons φ^I in Eq. (2.13) disappear from the expressions. Then the Yukawa terms are

$$\mathcal{L}_Y = -\frac{1}{\sqrt{2}} (v + s) \left(y^u q_1 \bar{u} + y^d q_2 \bar{d} + y^e \ell_2 \bar{e} + \text{c.c.} \right), \tag{2.18}$$

where the components of q and ℓ are evaluated in the unitary gauge. Combining the fields appearing in each term of Eq. (2.18) into Dirac spinors

$$\psi^u := \begin{pmatrix} q_1 \\ \bar{u}^\dagger \end{pmatrix}, \quad \psi^d := \begin{pmatrix} q_2 \\ \bar{d}^\dagger \end{pmatrix}, \quad \psi^v := \begin{pmatrix} \ell_1 \\ 0 \end{pmatrix}, \quad \psi^e := \begin{pmatrix} \ell_2 \\ \bar{e}^\dagger \end{pmatrix}, \tag{2.19}$$

$\mathcal{L}_Y$ defines the Dirac fermion masses and their interactions with the Higgs boson.

The SM Lagrangian reorganized to reflect electroweak symmetry breaking (EWSB) is thus written in terms of the mass eigenfields of the fermions,[1] Higgs boson, and gauge bosons. The fermions and Higgs mass eigenfields are each assigned a quantum number called isospin T_3, whose value is $+\frac{1}{2}$ for fields that are the first component of the SU(2) fundamental representation in the unitary gauge, $-\frac{1}{2}$ for those that are the second component, and 0 for those that are an SU(2) singlet. Returning to the example given in the beginning of this section, the neutrinos, which remain massless and have isospin $T_3 = \frac{1}{2}$, are now clearly distinct from charged leptons, which have now acquired mass as Dirac spinors.

The employment of the unitary gauge in Eq. (2.14) does not fix the U(1) gauge, implying that there is a U(1) gauge symmetry left after EWSB. In fact, gauge transformation of the form

$$B_\mu \to B_\mu + \cos \theta_{\mathrm{W}} \partial_\mu \chi, \tag{2.20}$$

$$W'^3_\mu \to W'^3_\mu + \sin \theta_{\mathrm{W}} \partial_\mu \chi, \tag{2.21}$$

where χ is any scalar function, leaves Z_μ unchanged, but transforms the orthogonal combination of B_μ and W'^3_μ,

$$A_\mu := \sin \theta_{\mathrm{W}} W'^3_\mu + \cos \theta_{\mathrm{W}} B_\mu, \tag{2.22}$$

as

$$A_\mu \to A_\mu + \partial_\mu \chi. \tag{2.23}$$

The vector potential A_μ, which represents the photon field, does not have a mass term and thus is the gauge boson of the remaining U(1) gauge interaction, which is the electromagnetic force. The electromagnetic charge Q, the phase scale factor of the new U(1) symmetry, can be calculated as

$$Q = T_3 + Y. \tag{2.24}$$

---

[1]The fermion mass matrices

$$M^f := \frac{v y^f}{\sqrt{2}} \quad f = u, d, e$$

can be diagonalized by biunitary transformations. This amounts to a change of basis in the three-dimensional generation space. While such a change can be rotated away by field redefinitions for the $\bar{u}$, $\bar{d}$, ℓ, and $\bar{e}$ fields, the charged current terms (2.26) imply that the change of basis is physical for q. In other words, there exist direct inter-generational interactions for ψ^u and ψ^d, encoded in the Cabibbo–Kobayashi–Maskawa (CKM) quark-mixing matrix U^{CKM}. Nevertheless, the CKM matrix is known to be close to identity, i.e., $|U^{\mathrm{CKM}}_{ii}| \gg |U^{\mathrm{CKM}}_{jk}|$ for $j \neq k$, such that generation, or equivalently flavor, mixing of quarks is possible but rare. The CKM matrix is also responsible for the so-called CP-violating interactions, where the rate of one process occurring is not equal to that of the CP-conjugate process. However, again due to the closeness of the CKM matrix to identity, CP violation is a small effect in the SM.

The electromagnetic interaction of fermions has the form

$$\mathcal{L}_{\text{EM}} = -QeA_\mu\overline{\psi}\gamma^\mu\psi, \tag{2.25}$$

where $\overline{\psi} := \psi^\dagger\gamma^0$ and $e := g_1\cos\theta_{\text{W}}$. The Dirac matrices γ^μ are given by $\gamma^\mu := \begin{pmatrix} & \overline{\sigma}^\mu \\ \sigma^\mu & \end{pmatrix}$.

The massive W and Z bosons, single-particle states of $W^\pm$ and Z fields, mediate the short-range weak interaction. The part of the Lagrangian containing the $W^\pm$ terms is called the charged current

$$\mathcal{L}_{\text{c.c.}} = -\frac{g_2}{\sqrt{2}}W_\mu^+\overline{\psi}'\gamma^\mu\frac{1}{2}(1-\gamma^5)\psi + \text{c.c.}, \tag{2.26}$$

where $\gamma^5 = i\gamma^0\gamma^1\gamma^2\gamma^3$. The charged current causes direct interactions between different mass eigenfields, also called particle flavors. Even after EWSB, the charged current interactions connect only the fermion fields in the doublet representation of SU(2), i.e., the left-handed components of the Dirac fermions, and is said to maximally violate parity. The operator $\frac{1}{2}(1-\gamma^5)$ projects the left-handed component out. The interaction with Z is called the neutral current and has the form

$$\mathcal{L}_{\text{n.c.}} = -\frac{g_2}{2\cos\theta_{\text{W}}}Z_\mu\overline{\psi}\gamma^\mu\frac{1}{2}(g_V - g_A\gamma^5)\psi, \tag{2.27}$$

where g_V and g_A are, respectively, called the vector and axial vector couplings. They can be calculated from the isospin and hypercharge of the respective fields.

2.3   Renormalizations in the Standard Model

A renormalizable QFT that permits perturbative calculations can only possess a finite number of divergent terms in its perturbative expansion. All of such divergences can then be absorbed by the input parameters of the Lagrangian, leaving the physical observables finite in all orders of perturbation. The SM is such a renormalizable theory. Its interaction Lagrangian consists only of terms where the mass dimension of the coupling constant is non-negative.

In the language of renormalization group (RG), a renormalizable Lagrangian is one that contains only relevant and marginal terms, or terms whose coefficients may grow as the length scale of the system increases. Accordingly, a non-renormalizable Lagrangian contains irrelevant terms, i.e., terms whose coefficients shrink as the length scale increases. However, the shrinking of the coefficients of the irrelevant terms implies that a theory that is non-renormalizable at a high-energy scale can appear renormalizable in a lower-energy scale. Conversely, a renormalizable theory can be an effective low-energy limit of some other fundamental theory, which might contain coupling constants with negative mass dimensions. In fact, the current general consensus is that the SM must be such an effective theory that is valid only up to some energy scale higher than the electroweak scale of $\mathcal{O}(100)$ GeV. This belief emerges from the existence of observations that are not explained by the SM, discussed in Sect. 2.5.

In the SM Lagrangian introduced in Sect. 2.1, gauge coupling constants $\{g_i\}$, Yukawa matrix elements $\{y_{ij}^X\}$, and the parameter λ in the Higgs potential are the coefficients of marginal terms. These parameters may or may not grow under RG evolution, or "running," from a high-energy scale to lower. The most consequential is the running of the strong coupling constant g_3 which, due to the specific field content of the SM, runs upward rapidly with decreasing energy scale. In other words, the SU(3) gauge interaction, called quantum chromodynamics (QCD), is an asymptotically free theory; it permits perturbative calculation in the high-energy regime but becomes strongly coupled below ~ 1 GeV. In contrast, the electromagnetic coupling constant $e = g_1 g_2 / \sqrt{g_1{}^2 + g_2{}^2}$ runs in the opposite direction, with the low-energy asymptotic value for $\alpha = e^2/4\pi$ of $\sim 1/137$.

The only relevant term, i.e., the term whose coefficient always grows when evolved down in energy scale, is the quadratic term of the Higgs potential. Its coefficient μ^2 in Eq. (2.10) thus is a peculiar parameter of the SM. The implication of the running of μ^2 is also discussed in Sect. 2.5.

2.4 Observed Particles

The basic observables of the SM are single-particle states of the various mass eigenfields and the rates of their interactions. To summarize the mass eigenfields, there are three generations each of ψ^u, ψ^d, ψ^ν, and ψ^e in Eq. (2.19), the Higgs boson, photon, gluon, and the weak bosons $W^\pm$ and Z. Of those, the neutrinos, the photon, and the gluon are massless in the SM, although it is experimentally well-established that at least two of the neutrinos have nonzero mass. The mass spectrum of the rest of the particles has a large dynamic range, spanning six orders of magnitude. The names and masses of SM particles are listed in Fig. 2.1 along with their interactions. Note that each of the leptons and quarks has an antiparticle with a sign-inverted electromagnetic charge. The naming convention of the antiparticles is to prefix the particle names with anti-, except for the electron whose antiparticle is called the positron. However, in most contexts, a mention of a particle implies its antiparticle as well.

The single-particle state of the photon (γ), historically known as gamma ray, appears in interactions involving electromagnetically charged particles. This includes not only the charged fermions but also the weak bosons $W^\pm$; the self-interaction term in the SU(2) field strength in Eq. (2.7) predicts interactions such as $W_\mu^+ A_\nu \partial^\mu W^{-\nu}$ after EWSB.

The charged leptons ψ^e have well-defined masses which are precisely measured. Their electromagnetic charge Q is -1. The first generation, i.e., the lightest in mass, is the electron (e) with a mass of 511 keV. It was the first elementary particle

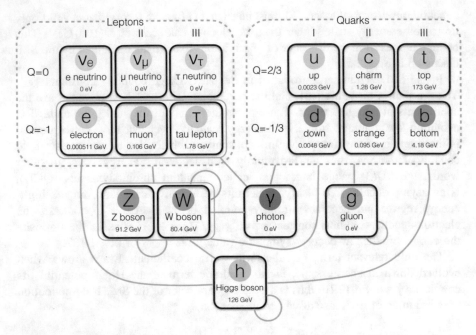

Fig. 2.1 The list of SM particles. Each round square represents a mass eigenfield of the SM, with its name and mass indicated at the *bottom*. Particles and particle groups connected by a *grey line* interact directly. Particles connected to themselves exhibit self-interactions

to be experimentally recognized, by the observation that the cathode ray consists of particles whose charge-to-mass ratio was far higher than any ions. The second generation, muon (μ), whose mass is 106 MeV, was found in cosmic rays in 1936, but was recognized as "a heavier version of the electron" only much later. The heaviest charged lepton is the tau lepton (τ) with a mass of 1.78 GeV.

The neutrinos ψ^{ν}, being mass degenerate in the SM, are labeled by the leptons that they accompany in the charged current interaction defined in Eq. (2.26). In fact, being electromagnetically neutral, the charged and neutral currents are the only interactions neutrinos exhibit, which makes them extremely elusive. The absence of the Yukawa terms for neutrinos in the SM implies lepton number conservation, i.e., the absence of inter-generational interactions between charged lepton and neutrino flavors. However, neutrinos are observed to oscillate from one flavor to another, indicating clearly that the SM is not a complete theory even of the known particles. If neutrino oscillation is due to their Yukawa interactions with the Higgs boson, quantum number counting immediately shows that there must be a new G_{SM}-singlet (representation $(\mathbf{1}, \mathbf{1}, 0)$) fermion field in the theory.

The three generations of up-type quarks ψ^{u} are named up (u), charm (c), and top (t) from the lightest to heaviest. The names of down-type quarks ψ^{d} are down (d), strange (s), and bottom (b). A barred quark symbol such as $\bar{t}$ indicates the antiparticle of the quark. The quarks and the gluon are charged under the asymptotically free color SU(3). The strengthening of the QCD interaction at long distances results in the so-called color confinement, or the formation of color-singlet states. In other words, quarks and gluons do not exist in single-particle asymptotic states, but rather form compound states with other quarks and gluons, collectively called hadrons. Hadron single-particle states exist only when the particle combinations are SU(3) singlets. Most of the known states are understood to consist of three quarks or a quark and an antiquark, corresponding to the singlets in the representation decomposition $\mathbf{3} \otimes \mathbf{3} \otimes \mathbf{3} = \mathbf{10} \oplus \mathbf{8} \oplus \mathbf{8} \oplus \mathbf{1}$ and $\mathbf{3} \otimes \bar{\mathbf{3}} = \mathbf{8} \oplus \mathbf{1}$. The two types of hadrons are, respectively, called baryons and mesons.

It should be noted that most of the single-particle states are not truly asymptotic and should be described as resonances. Resonances decay, i.e., transition to specific multi-particle states of fields with lower masses. The existence of a resonance is in fact often inferred from the invariant mass and particle content of the multi-particle state. Due to the uncertainty principle, this invariant mass has a finite distribution, the center of which is the nominal mass of the resonance. The only particles whose decays have not been observed are the photon, the electron, the neutrinos, and the proton (p), which is a baryon with flavor content of uud. Many atomic nuclei, which are bound states of protons and neutrons (n, udd), are also stable. However, the neutron itself decays with a mean lifetime of 882 s when isolated.

Single-particle state of the Higgs boson is a rapidly decaying resonance. The Higgs boson was discovered in 2012 [1, 2] and has a mass of 126 GeV. This mass is too light for on-shell decays, or decays to daughters with their nominal masses, to two weak bosons as dictated by Eq. (2.15). Therefore, the decay branching ratio (BR) of the Higgs boson is not dominated by W^+W^- and ZZ final states, as it would be if the Higgs boson was heavier than $\sim$180 GeV, but by $b\bar{b}$. The bottom quark decay channel dominates because, as seen in Eq. (2.18), the fermion mass is directly proportional to its Yukawa coupling to the Higgs field, and the bottom quark is the heaviest fermion that the Higgs boson can decay to. There are nevertheless non-negligible branchings to off-shell W^+W^-, ZZ, and also to $\gamma\gamma$, the latter being possible due to higher-order couplings. It was in the ZZ and $\gamma\gamma$ final states that the Higgs boson discovery was first established.

The heaviest fermion, and thus the one that couples to the Higgs boson most strongly, is actually the top quark. However, its mass of 173 GeV is heavier than the Higgs mass, and therefore even the off-shell decay, while in principle possible, is practically negligible. The heaviness of the top quark has the interesting consequence that it decays before forming a hadron, mostly to W^+ and b.

The weak bosons $W^{\pm}$ and Z themselves rapidly decay into fermions. Since weak interaction is universal, i.e., has strength fixed by the single parameter g_2, the decay BR of the W boson can be calculated simply by the fermion masses and their SU(3) representations to good approximation. The decay branchings of the Z boson is similarly obtained, but by additionally using the hypercharge of the fermions. The measured Z resonance width, which is related to the number of fermions it couples to, was used to constrain the number of light neutrino generations, which is difficult to probe otherwise.

Resonances with lifetimes longer than $\mathcal{O}(10)$ ns are considered stable from the experimental point of view if the size of the apparatus is smaller than $\mathcal{O}(10)$ m. These so-called collider-stable particles include muons and many hadrons such as n, π^+ (charged pion, $u\bar{d}$), K^+ (charged kaon, $u\bar{s}$), K^0_L (K long, $\frac{1}{\sqrt{2}}(d\bar{s} + \bar{d}s)$), and their antiparticles.

2.5   Unanswered Questions in the Standard Model

The Standard Model, despite its remarkable success, does not explain certain observational facts [3], as mentioned in Chap. 1. There are also features inherent in the SM which, while posing no contradiction to observation, may be nevertheless philosophically unpleasing.

The list of the major observational facts not incorporated in the SM includes neutrino oscillation, existence of dark matter, and the matter–antimatter imbalance in the universe. The first has already been mentioned and implies that there should be terms in the SM Lagrangian that describe neutrino masses, possibly involving a new singlet field.

The second comes from astronomical observations about the existence of gravitationally interacting entities in the universe whose electromagnetic and strong interactions are severely suppressed or nonexistent. It is well established that the neutrinos of the SM can only account for a small fraction of this energy-mass that is the backbone of the large-scale structure of the universe. The remaining unexplained mass is collectively called dark matter. It is not clear that dark matter permits a QFT description, but if it does, then there must be new fields in the Lagrangian beyond those listed in Table 2.1.

The last issue, the dominance of matter over antimatter in the universe, requires a large CP-violation effect in the Lagrangian. The only source of CP violation in the SM, which proves to be insufficient to explain the matter–antimatter asymmetry in the universe, is in the quark mixing. The neutrino mass terms in fact might provide additional CP violation, but it is not clear how large the effect would be.

Even in the absence of such observations, not being able to explain why the 18 parameters of the SM take the values they have is not quite satisfying. One can also ask why the gauge group of the SM is $SU(3) \times SU(2) \times U(1)$, why there are three generations of quarks and leptons, etc.; it is possible to demand that all features of the SM be explained as a consequence of some dynamics with a minimal input. Although such an extreme position is not common, at least some of the features are often regarded as immediate problems. An example is the so-called strong CP problem [4]: The SM can in fact harbor 19 parameters, the extra one being the coefficient of the CP-violating Yang-Mills curvature of the gluon field, but non-observation of CP violation in the strong sector states that the parameter is consistent with zero, i.e., appears finely tuned to be extremely small. Another example, also associated with fine-tuning, is the infamous UV-sensitivity of the Higgs mass, discussed below.

Mass terms, or parts of the potential that becomes quadratic in the equation of motion of the fields, evolve with the RG flow in interacting QFT. Even if the mass term is absent for a field in the Lagrangian at the input scale, one emerges at low-energy scales unless it is forbidden by some symmetry. In the SM, mass terms for the fermions and the vector bosons are disallowed by gauge symmetries. In the absence of the Yukawa terms, fermion masses are further protected by chiral symmetry. There is, however, no such protection for the Higgs quadratic coefficient μ^2.

Unprotected mass depends directly and inhomogenously on the UV cutoff scale Λ_{UV}. From a simple dimensional analysis, then, μ^2 should receive a contribution proportional to Λ_{UV}^2 when evolved down with RG. This quadratic dependency can be worrisome, because the mass $m_H \sim 126 \, \text{GeV}$ and the VEV $v = 246 \, \text{GeV}$ of the Higgs field imply that μ^2 in the input scale has a value finely tuned to cancel this additive RG evolution. If Λ_{UV} is at the Planck-scale $M_{Pl} \sim 10^{19} \, \text{GeV}$, then the fine-tuning must be done over thirty orders of magnitude.

The problem of fine-tuning exists because the Higgs field is assumed to be a fundamental scalar appearing in the Lagrangian at the UV cutoff. There are multiple possible approaches to resolve the issue. One would be to not regard it as a problem to begin with; accepting an anthropic principle, fine-tuning may be a necessity for a sentient being to emerge and observe the universe. Another approach is to postulate that the Higgs field at the electroweak scale is actually a compound, typically constituting of two fermions coupled by some asymptotically free interaction. In such scenarios, the Higgs boson is resolved above some energy scale that can be close to the electroweak scale, requiring no fine-tuning at the fundamental scale. Yet another possibility is to retain the fundamental scalar Higgs field by introducing new symmetries that eliminate the Λ_{UV}^2 dependence of μ^2. For example, if the Higgs boson is the pseudo-Nambu–Goldstone boson of some spontaneously broken approximate symmetry, as in the Little Higgs Model [5], its mass scale is determined by the parameters related to this approximate symmetry and not by Λ_{UV}.

Supersymmetry is another symmetry that has the effect of controlling the scalar mass. In fact, a supersymmetric Lagrangian forbids quadratic running of any scalar mass, not just the Higgs field. This is achieved because supersymmetry demands that all scalars are mass degenerate to some fermions. Since fermion mass parameters run at most logarithmically with Λ_{UV}, the quadratic dependence automatically disappears. The next sections describe the principle of supersymmetry and the supersymmetric extension of the SM.

2.6 Supersymmetry

Supersymmetry (SUSY) [6–9] is a continuous symmetry that induces transformations between bosons and fermions. Every field in a supersymmetric Lagrangian is given a set of bosonic and fermionic degrees of freedom combined in a so-called supermultiplet. A supersymmetry transformation is represented on the space of supermutiplets, i.e., it mixes the components of the multiplets. The Lagrangian is formed to be invariant under such a transformation.

An implementation of SUSY is given in superspace, or regular four-dimensional spacetime augmented with two complex Grassmann dimensions θ^1 and θ^2. Functions over superspace, called superfields, embody the supermultiplets mentioned above. A series expansion of a general superfield $S(x, \theta, \theta^\dagger)$, where x is the regular spacetime coordinate, in powers of Grassmann variables is given as

$$S(x, \theta, \theta^\dagger) = a + \theta\xi + \theta^\dagger\chi^\dagger + \theta\theta b + \theta^\dagger\theta^\dagger c + \theta^\dagger\overline{\sigma}^\mu\theta v_\mu + \theta^\dagger\theta^\dagger\theta\eta + \theta\theta\theta^\dagger\lambda^\dagger + \theta\theta\theta^\dagger\theta^\dagger d.$$
(2.28)

Scalars a, b, c, and d, vector v, and spinors ξ, $\chi^\dagger$, η, and $\lambda^\dagger$ are all functions of x and form a supermultiplet. Note that for a complex S, there are 2 bosonic degrees of freedom in each of the scalars and 8 in v, while each spinor has 4 fermionic degrees of freedom. For a real S, $a^* = a$, $c = b^*$, $d^* = d$, $v^* = v$, $\chi = \xi$, and $\lambda = \eta$, therefore there are 8 degrees of freedom for both bosons and fermions. In short, there are equal numbers of bosons and fermions for any given superfield.

The SUSY transformation of superfields is generated by two differential operators Q_a and $Q_{\dot{a}}^\dagger$:

$$Q_a := i\frac{\partial}{\partial\theta^a} - \frac{1}{2}(\sigma^\mu\theta^\dagger)_a\partial_\mu,$$

$$Q_{\dot{a}}^\dagger := -i\frac{\partial}{\partial\theta^{\dagger\dot{a}}} + \frac{1}{2}(\theta\sigma^\mu)_{\dot{a}}\partial_\mu.$$
(2.29)

Since the operators carry spinor indices, the transformation also requires a spinor parameter. The shift $\delta_\zeta S$ of S under SUSY with parameter ζ is

$$\delta_\zeta S := -i(\zeta Q + Q^\dagger\zeta^\dagger)S = S\left(x^\mu - \frac{i}{2}\theta\sigma^\mu\zeta^\dagger - \frac{i}{2}\theta^\dagger\overline{\sigma}^\mu\zeta, \theta + \zeta, \theta^\dagger + \zeta^\dagger\right) - S(x^\mu, \theta, \theta^\dagger)$$
(2.30)

up to the first power in ζ. Thus, the SUSY transformation of superfields is realized as a rotation in $x - (\theta, \theta^\dagger)$ space and a translation in the θ and $\theta^\dagger$ direction. The rotation appears as a translation if projected onto the x direction, i.e., SUSY shifts conventional functions in the regular spacetime. The relation between the SUSY transformation and a conventional translation can also be found in the anticommutator of the SUSY operators:

$$\{Q_a, Q_b\} = \{Q_{\dot{a}}^\dagger, Q_{\dot{b}}^\dagger\} = 0,$$
(2.31)

$$\{Q_a, Q_{\dot{a}}^\dagger\} = i(\sigma^\mu)_{a\dot{a}}\partial_\mu = (\sigma^\mu)_{a\dot{a}}P_\mu,$$
(2.32)

where $P_\mu = i\partial_\mu$ is the differential representation of the momentum operator. Equation (2.32) in fact hints that SUSY is a part of the spacetime symmetry that extends the Poincaré group. Indeed, the commutation relations of Q and $Q^\dagger$ with P_μ and the Lorentz transformation $M_{\mu\nu}$ are

$$[Q_a, P_\mu] = [Q_{\dot{a}}^\dagger, P_\mu] = 0,$$
(2.33)

$$[\mathcal{Q}_a, \mathcal{M}_{\mu\nu}] = \frac{i}{4}(\sigma_\mu\bar{\sigma}_\nu - \sigma_\nu\bar{\sigma}_\mu)_a{}^b\mathcal{Q}_b, \tag{2.34}$$

$$[\mathcal{Q}_{\dot{a}}^\dagger, \mathcal{M}_{\mu\nu}] = -\frac{i}{4}\mathcal{Q}_{\dot{b}}^\dagger(\bar{\sigma}_\mu\sigma_\nu - \bar{\sigma}_\nu\sigma_\mu)_{\dot{a}}^{\dot{b}}, \tag{2.35}$$

which indicates that Poincaré and SUSY generators form a closed algebra.

There is a SUSY-invariant subspace of the space of superfields which gets used extensively in the supersymmetric extension of the SM. The subspace is defined as a kernel of a differential operator

$$\mathcal{D}_{\dot{a}}^\dagger = \frac{\partial}{\partial\theta^{\dagger\dot{a}}} - \frac{i}{2}(\theta\sigma^\mu)_{\dot{a}}\partial_\mu. \tag{2.36}$$

The superfield Φ is called a left-chiral superfield if it satisfies $\mathcal{D}^\dagger\Phi = 0$. One can show that

$$\{\mathcal{D}_{\dot{a}}^\dagger, \mathcal{Q}_a\} = \{\mathcal{D}_{\dot{a}}^\dagger, \mathcal{Q}_{\dot{b}}^\dagger\} = 0, \tag{2.37}$$

and thus any SUSY transform of a left-chiral superfield is also a left-chiral superfield. The hermitian conjugate of $\mathcal{D}^\dagger$,

$$\mathcal{D}_a = \frac{\partial}{\partial\theta^a} - \frac{i}{2}(\sigma^\mu\theta^\dagger)_a\partial_\mu, \tag{2.38}$$

annihilates right-chiral superfields, which are the complex conjugates of left-chiral superfields.

The expression of left-chiral superfields can be simplified with a change of coordinates

$$y^\mu := x^\mu - \frac{i}{2}\theta\sigma^\mu\theta^\dagger, \quad \rho := \theta. \tag{2.39}$$

In the new variables y and ρ, the operator $\mathcal{D}^\dagger$ is

$$\mathcal{D}_{\dot{a}}^\dagger = \frac{\partial}{\partial\rho^{\dagger\dot{a}}}. \tag{2.40}$$

Therefore, a left-chiral superfield is a function only of y and ρ. Expanding Φ in powers of ρ,

$$\Phi(y, \rho) = \phi(y) + \rho\psi(y) + \frac{1}{2}\rho\rho F(y). \tag{2.41}$$

The operators $\mathcal{Q}$ and $\mathcal{Q}^\dagger$ expressed in the new variables are

$$Q_a = i\frac{\partial}{\partial \rho^a},$$

$$Q_{\dot{a}}^{\dagger} = -i\frac{\partial}{\partial \rho^{\dagger \dot{a}}} + (\rho \sigma^{\mu})_{\dot{a}}\frac{\partial}{\partial y^{\mu}}. \tag{2.42}$$

From Eqs. (2.41) and (2.42), the SUSY transformations of the component fields ϕ, ψ, and F are given as

$$\delta_{\zeta}\phi = \zeta \psi, \tag{2.43}$$

$$\delta_{\zeta}\psi_a = -i(\sigma^{\mu}\zeta^{\dagger})_a \partial_{\mu}\phi + \zeta_a F, \tag{2.44}$$

$$\delta_{\zeta}F = -i\zeta^{\dagger}\overline{\sigma}^{\mu}\partial_{\mu}\psi. \tag{2.45}$$

Equation (2.41) shows that there are only four bosonic and fermionic degrees of freedom for a left-chiral superfield. Re-expression of Eq. (2.41) with x and θ results in

$$\Phi(x, \theta, \theta^{\dagger}) = \phi(x) + \theta \psi(x) + \frac{1}{2}\theta\theta F(x)$$

$$+ \frac{i}{2}\theta^{\dagger}\overline{\sigma}^{\mu}\theta \partial_{\mu}\phi(x) - \frac{i}{4}\theta\theta\theta^{\dagger}\overline{\sigma}^{\mu}\partial_{\mu}\psi(x) - \frac{1}{16}\theta\theta\theta^{\dagger}\theta^{\dagger}\partial^2\phi(x). \tag{2.46}$$

The term in the left-chiral superfields proportional to $\frac{1}{2}\theta\theta$ is called an F-term. Equation (2.45) shows that the coefficient field of the F-term is shifted by a total derivative under the SUSY transformation. Therefore, the x-integral of the F-term is invariant under SUSY.

Another interesting feature of the expansion in Eq. (2.46) can be seen by integrating $\Phi^*\Phi$ over the Grassmann dimensions:

$$\int d^2\theta d^2\theta^{\dagger}\ \Phi^*\Phi = \partial^{\mu}\phi^*(x)\partial_{\mu}\phi(x) + i\psi^{\dagger}(x)\overline{\sigma}^{\mu}\partial_{\mu}\psi(x) + F^*(x)F(x), \tag{2.47}$$

where integration measures are defined as $d^2\theta := d\theta^1 d\theta^2$ and $d^2\theta^{\dagger} := d\theta^{\dagger 2}d\theta^{\dagger 1}$. The right-hand side of Eq. (2.47) supplies the quadratic terms of ϕ, ψ, and F. For the first two fields, these are the kinetic terms, while for F there is no propagation. For this reason, F is called an auxiliary field. The integration isolates the terms proportional to $\frac{1}{4}\theta\theta\theta^{\dagger}\theta^{\dagger}$, which are called D-terms. In general, D-terms are SUSY-invariant up to total derivatives. This is evident from Eq. (2.29) since only the second terms on the right-hand side contribute to the SUSY transformation of the D-term.

Therefore, a SUSY-invariant conventional Lagrangian can be constructed from left-chiral superfields $\{\Phi_i\}$ by taking the D-term of $\Phi_i^*\Phi_i$ and the F-terms of various products $\Phi_i\Phi_j \cdots$ and their complex conjugates. A product of left-chiral superfields is also a left-chiral superfield since the operator $\mathcal{D}^{\dagger}$ follows the Leibniz rule.

The sum of the products $\Phi_i\Phi_j\cdots$ is called a superpotential and is often denoted by W. Thus, symbolically, a SUSY-invariant Lagrangian is given by

$$\mathcal{L} = \mathcal{L}_K + \mathcal{L}_W$$
$$= [\Phi_i^*\Phi_i]_D + \{[W(\Phi)]_F + \text{c.c.}\}, \qquad (2.48)$$

where $[\cdot]_D$ and $[\cdot]_F$ stand for extractions of D- and F-terms.

When the scalars have mass dimension 1, the most general renormalizable superpotential has the form

$$W = l^i\Phi_i + \frac{1}{2}m^{ij}\Phi_i\Phi_j + \frac{1}{6}y^{ijk}\Phi_i\Phi_j\Phi_k, \qquad (2.49)$$

where parameters m^{ij} and y^{ijk} are symmetric under permutations of indices.

If gauge symmetry is imposed on W in Eq. (2.49), the combinations of the fields must be such that each term on the right-hand side is independently gauge invariant. In particular, the parameter l^i can be nonzero only if Φ_i is a singlet under all gauge symmetries of the theory. Gauge transformation of the left- and right-chiral superfields Φ and Φ^* is given by

$$\Phi \rightarrow \exp(2igT^I\Omega^I)\Phi,$$
$$\Phi^* \rightarrow \exp(-2igT^I\Omega^{I*})\Phi^*, \qquad (2.50)$$

where g is the coupling constant and Ω^I is a dimensionless chiral superfield coupled to the gauge group generator T^I. The generators follow the relations

$$[T^I, T^J] = if^{IJK}T^K, \qquad (2.51)$$

$$2\text{Tr}(T^IT^J) = \delta^{IJ} \qquad (2.52)$$

with some structure constant f^{IJK}. The kinetic terms in the Lagrangian in Eq. (2.48) must be modified to be gauge invariant. Dimensionless real superfields $\mathcal{V}^I$ can be inserted to absorb the transformations in Φ and Φ^* as

$$\mathcal{L}_K = [\Phi_i^* e^{2gT^I\mathcal{V}^I}\Phi_i]_D, \qquad (2.53)$$

if $\mathcal{V}$ transforms as

$$e^{2g\mathcal{V}} \rightarrow e^{2ig\Omega^\dagger}e^{2g\mathcal{V}}e^{-2ig\Omega}, \qquad (2.54)$$

where $\mathcal{V} := T^I\mathcal{V}^I$ and $\Omega := T^I\Omega^I$.

The superfield $\mathcal{V}^I$ is called a vector superfield and in general has eight bosonic and fermionic degrees of freedom, as counted at the beginning of this section. However, the supergauge transformation in Eq. (2.54) reveals that four of each of the bosonic and fermionic degrees of freedom are nonphysical. The so-called Wess–Zumino gauge partially fixes these gauge degrees of freedom, so that $\mathcal{V}^I$ has the form

$$\mathcal{V}^I|_{\text{W-Z}} = \frac{1}{2}\theta\sigma^\mu\theta^\dagger A_\mu^I + \frac{1}{2\sqrt{2}}(\theta^\dagger\theta^\dagger\theta\lambda^I + \text{c.c.}) - \frac{1}{8}\theta\theta\theta^\dagger\theta^\dagger D^I. \tag{2.55}$$

The real vector field A_μ^I is the conventional gauge vector potential, the spinor λ^I is the corresponding gaugino field, and the real scalar D^I is an auxiliary field. Using the Wess–Zumino gauge, one can obtain the conventional gauge-invariant kinetic and additional gauge interaction terms from Eq. (2.53):

$$\mathcal{L}_{\text{K}} = (\nabla^\mu\phi_i)^\dagger(\nabla_\mu\phi_i) + i\psi_i^\dagger\bar{\sigma}^\mu\nabla_\mu\psi_i + F_i^*F_i - \sqrt{2}g(\phi_i^*T^I\psi_i\lambda^I + \text{c.c.}) - g\phi_i^*T^I\phi_iD^I. \tag{2.56}$$

The kinetic terms of A_μ^I and λ^I are obtained from the left-chiral superfield

$$S_a := T^I S_a^I = T^I\frac{1}{4g}\left[\mathcal{D}^\dagger\mathcal{D}^\dagger\left(e^{-2g\mathcal{V}}\mathcal{D}_a e^{2g\mathcal{V}}\right)\right]^I. \tag{2.57}$$

Here, $\mathcal{D}^{\dagger\dot{a}}$ is defined to satisfy $\mathcal{D}_{\dot{a}}^\dagger\xi^{\dot{a}} = \mathcal{D}^{\dagger\dot{a}}\xi_{\dot{a}}$. In the Wess–Zumino gauge, S^I reduces to

$$S^I|_{\text{W-Z}}(y,\rho) = \frac{1}{\sqrt{2}}\lambda^I(y) - \frac{1}{2}\rho D^I(y) - \frac{1}{2}\sigma^{\mu\nu}\rho F_{\mu\nu}^I(y) + \frac{i}{2\sqrt{2}}\rho\rho\sigma^\mu(\nabla_\mu\lambda)^{I\dagger}(y), \tag{2.58}$$

where

$$F_{\mu\nu}^I = \partial_\mu A_\nu^I - \partial_\nu A_\mu^I - gf^{IJK}A_\mu^J A_\nu^K \tag{2.59}$$

is the field-strength tensor, and

$$(\nabla_\mu\lambda)^I = \partial_\mu\lambda^I - gf^{IJK}A_\mu^J\lambda^K \tag{2.60}$$

is the gauge-covariant derivative of the gaugino field. The F-term of $S^I S^I$ then yields

$$[S^I S^I]_F = i\lambda^I\sigma^\mu(\nabla_\mu\lambda)^{I\dagger} + \frac{1}{2}D^I D^I - \frac{1}{4}F^{I\mu\nu}F_{\mu\nu}^I - \frac{1}{4}F^{*I\mu\nu}F_{\mu\nu}^I, \tag{2.61}$$

where $F^{*I\mu\nu} := \frac{1}{2}\epsilon^{\mu\nu\alpha\beta}F_{\alpha\beta}^I$.

The discussion above is valid for a single vector superfield. When the gauge group is a direct product of subgroups, as in the SM, each subgroup has a vector supermultiplet. Enumerating the groups with index A, the Lagrangian (2.48) of the supersymmetric gauge theory should be augmented with

$$\mathcal{L}_{\text{SuperYM}} = [2\text{Tr}(S_A S_A)]_F. \tag{2.62}$$

It is apparent from Eqs. (2.56) and (2.61) that the auxiliary fields F_i and D_A^I do not propagate and only have quadratic and linear terms. Therefore these fields can be easily integrated out, effectively resulting in the substitutions

$$F_i \to -W_i^* := -\left.\frac{\delta W^*}{\delta \Phi_i^*}\right|_{\Phi_i^* \to \phi_i^*}, \tag{2.63}$$

$$F_i^* \to -W_i := -\left.\frac{\delta W}{\delta \Phi_i}\right|_{\Phi_i \to \phi_i}, \tag{2.64}$$

$$D_A^I \to g_A \phi_i^* T_A^I \phi_i. \tag{2.65}$$

Here, $|_{\Phi_i \to \phi_i}$ indicates replacing the left-chiral superfields in the expression to the left by their scalar components. With these replacements, the full form of the Lagrangian is

$$\begin{aligned}
\mathcal{L}_{\text{SUSY}} &= \mathcal{L}_{\text{K}} + \mathcal{L}_{\text{W}} + \mathcal{L}_{\text{SuperYM}} \\
&= [\Phi_i^* e^{2g_A V_A} \Phi_i]_D + \{[W(\Phi)]_F + \text{c.c.}\} + [2\text{Tr}(S_A S_A)]_F \\
&= (\nabla^\mu \phi_i)^\dagger (\nabla_\mu \phi_i) + i\psi_i^\dagger \overline{\sigma}^\mu \nabla_\mu \psi_i - V(\phi, \phi^*) - \frac{1}{2}(W_{ij}\psi_i\psi_j + \text{c.c.}) \\
&\quad - \frac{1}{4}F_A^{I\mu\nu}F_{A\mu\nu}^I - \frac{1}{4}F_A^{*I\mu\nu}F_{A\mu\nu}^I + i\lambda_A^{I\dagger}\overline{\sigma}^\mu \nabla_\mu \lambda_A^I - \sqrt{2}g_A(\phi_i^* \lambda_A \psi_i + \text{c.c.}),
\end{aligned} \tag{2.66}$$

where

$$V(\phi, \phi^*) := W_i^* W_i + g_A^2(\phi_i^* T_A^I \phi_i)^2, \tag{2.67}$$

$$W_{ij} := \left.\frac{\delta^2 W}{\delta\Phi_i \delta\Phi_j}\right|_{\Phi_i \to \phi_i}, \tag{2.68}$$

and

$$\lambda_A := T_A^I \lambda_A^I. \tag{2.69}$$

In Eq. (2.69), summation over A is not implied on the right-hand side.

As a conclusion to this brief summary of the principles of supersymmetry, the relation in scalar and fermion masses mentioned in Sect. 2.5 can be proven. In general, superpotential in Eq. (2.49) can be further extended with non-renormalizable terms. However, the mass term in the scalar potential given in Eq. (2.67) can only come from quadratic terms:

$$\mathcal{L}_{\text{scalar mass}} = -m^{*ij}m^{jk}\phi_i^* \phi_k. \tag{2.70}$$

Similarly, the fermion mass terms in Eq. (2.66) are due to constant terms in Eq. (2.68):

$$\mathcal{L}_{\text{fermion mass}} = -\frac{1}{2}m^{ij}\psi_i\psi_j - \frac{1}{2}m^{*ij}\psi_i^\dagger\psi_j^\dagger. \tag{2.71}$$

The equations of motion are then

$$\partial^2\phi_i + m^{*ij}m^{jk}\phi_k + \cdots = 0, \tag{2.72}$$

$$i\bar{\sigma}^\mu\partial_\mu\psi_i - m^{*ij}\psi_j^\dagger + \cdots = 0, \quad i\sigma^\mu\partial_\mu\psi_i^\dagger - m^{ij}\psi_j + \cdots = 0, \tag{2.73}$$

where the two equations in the second line can be combined into

$$\partial^2\psi_i + m^{*ij}m^{jk}\psi_k + \cdots = 0. \tag{2.74}$$

Therefore, the scalars and fermions of a supersymmetric Lagrangian have identical masses obtained by diagonalizing m^*m.

2.7   Supersymmetry Breaking

The mass degeneracy of the scalars and fermions in SUSY is actually problematic when considering the supersymmetric extension of the SM, since none of the SM fields are known to have partners of identical mass and different spin. Or, more to the point, bosons and fermions are distinguishable in our world, while in a supersymmetric universe there will only be superfields. Thus, to consider SUSY as a remedy to the problem of the fine-tuning of the Higgs mass, a non-exact form of SUSY which still leaves the scalar mass insensitive to Λ_{UV} must be known.

The next step is then to consider spontaneously broken SUSY. There, the scalar mass might not be protected any more by its degeneracy to a fermion, but the correction is bounded by at most some intermediate scale at which the effect of spontaneous breaking is resolved and exact SUSY is recovered. With this in mind, the supersymmetric standard model can be described by a low-energy effective Lagrangian in which the result of spontaneous SUSY breaking is parametrized by renormalizable terms that explicitly break the symmetry.

This effective approach postpones the question of how exactly SUSY is broken. It can actually be shown using Eq. (2.32) that SUSY is broken if $\langle V(\phi)\rangle \neq 0$. From Eqs. (2.63)–(2.65) and (2.67) this implies either or both of $\langle F_i\rangle \neq 0$ and $\langle D_A^I\rangle \neq 0$ for some i, A, I. The supersymmetric extension of the SM, however, does not have candidates of auxiliary fields with nonzero VEV. There is also a reason, discussed in Sect. 2.8, that the SUSY-breaking field cannot be one of the supersymmetrized SM fields. Therefore, from a phenomenological perspective, it is much more practical to study a general effective Lagrangian and its implications.

The form of the SUSY-breaking terms can be inferred from the constraint that the underlying dynamics respect SUSY. The implication of the constraint is that the fundamental Lagrangian also has the form in the second row of Eq. (2.66). The effective Lagrangian must come out of this fundamental Lagrangian as some fields acquire VEVs. It is then possible to write down an effective form of the fundamental Lagrangian using spurions, which are in this case superfields whose scalar components are replaced by constants, in such a way that the replacements result in a renormalizable Lagrangian.

The fundamental Lagrangian in the current analysis is

$$\mathcal{L}_{sp} = \left[Z^i \Phi_i^* e^{2\mathcal{V}'} \Phi_i \right]_D + \left\{ \left[\frac{1}{2} M^{ij} \Phi_i \Phi_j + \frac{1}{6} Y^{ijk} \Phi_i \Phi_j \Phi_k \right]_F + \text{c.c.} \right\} + \sum_A \left[2\gamma^A \text{Tr}(S_A' S_A') \right]_F,$$
(2.75)

where $\mathcal{V}' := g_A V_A$ and $S_A' := g_A S_A$, summation over A not implied. The term linear in Φ was dropped from the superpotential because there is no gauge-singlet field in the SM. Of the spurions in Eq. (2.75),

$$Z^i = 1 + \frac{1}{2}(z_1^i \theta\theta + \text{c.c.}) + \frac{1}{4} z_2^i \theta\theta\theta^\dagger\theta^\dagger$$
(2.76)

is real, and

$$M^{ij} = m^{ij} + \frac{1}{2} m_1^{ij}\theta\theta$$
(2.77)

$$Y^{ijk} = y^{ijk} + \frac{1}{2} y_1^{ijk}\theta\theta$$
(2.78)

$$\gamma^A = \frac{1}{g_A^2} + \frac{1}{2}\gamma_1^A\theta\theta$$
(2.79)

are left-chiral. Expanding the spurions,

$$\mathcal{L}_{sp} = \mathcal{L}_K + \mathcal{L}_W + \mathcal{L}_{SuperYM}$$
$$+ (z_1^i F_i^* \phi_i + \text{c.c.}) + z_2^i \phi_i^* \phi_i + \left(\frac{1}{2} m_1^{ij}\phi_i\phi_j + \frac{1}{6} y_1^{ijk}\phi_i\phi_j\phi_k + \text{c.c.} \right)$$
$$+ \sum_A \frac{1}{2}\gamma_1^A \lambda_A^I \lambda_A^I.$$
(2.80)

Because of the z_1^i factors, the substitution of auxiliary fields is modified. However, the substitutions produce terms of the form already present in Eq. (2.80). In summary, one finds that there are four types of SUSY-breaking terms: gaugino mass; scalar trilinear; scalar bilinear; and scalar mass. As expected, the effective SUSY-breaking terms introduced additional masses to gauginos and scalars, lifting the mass degeneracies of the vector and chiral supermultiplets, respectively. The bosonic and fermionic components in the supermultiplets are now physically distinguishable fields. They are said to be superpartners of each other.

It is worth noting that no dimensionless parameter appears in the SUSY-breaking terms. This has the important consequence that none of the masses in the theory gets a correction proportional to Λ_{UV} even in the presence of SUSY breaking. For this reason, SUSY is said to be broken softly when spontaneously broken. Assuming that the dimensionful parameters arise from some common dynamics, the typical mass scale of the parameters is denoted as M_{soft}. For a light scalar to be natural, $M_{soft} \sim \Lambda_{UV}$ is needed.

One mechanism of keeping M_{soft} low is to require SUSY breaking in a sector that is not directly coupled to the low-energy degrees of freedom. The effect of breaking should be communicated by an interaction common to the two sectors. Denoting the SUSY-breaking scale as $\sqrt{\langle F \rangle}$ and the communication strength as $1/M_{mess.}$,

$$M_{soft} \sim \frac{\langle F \rangle}{M_{mess.}}. \tag{2.81}$$

Thus, this indirect SUSY-breaking scenario allows more room for the theory to provide a desirable M_{soft}.

2.8 The Minimal Supersymmetric Standard Model

To seek a solution to the fine-tuning problem of the Higgs mass, it is appropriate to extend the SM with SUSY minimally, i.e., with a field content as similar as possible to that of the SM. Since only two types of superfields, left-chiral and vector, appeared in the analysis in Sect. 2.6, it follows that the SM fermions and the Higgs are elevated to chiral superfields and the gauge bosons to vector superfields.

Of the terms in Eq. (2.1), $\mathcal{L}_{kin.}$ and $\mathcal{L}_{YM}$ derive from $\mathcal{L}_K$ and $\mathcal{L}_{SuperYM}$ in Eq. (2.66). The remaining $\mathcal{L}_Y$ and $\mathcal{L}_V$ must then result from a superpotential. This association immediately faces a problem in Eq. (2.4); the terms $h \circ q\bar{u}$ and $h^* q\bar{d}$ cannot coexist in superpotential-based interactions. The minimal supersymmetric standard model (MSSM) thus needs another superfield whose scalar component replaces h^*. This new field is denoted as h_d since it couples to down-type quarks, and the h in the SM is renamed as h_u. The G_{SM}-representation of h_d is $(\mathbf{1}, \mathbf{2}, -\frac{1}{2})$.

The names of the new fields introduced by this extension are derived from the corresponding SM fields. Scalar partners of q, $\bar{u}$, and $\bar{d}$ are collectively called squarks. Similarly, scalar partners of ℓ and $\bar{e}$ are called sleptons. Squarks and sleptons together are sometimes referred to as sfermions. The fermionic partners of h_u and h_d are both called higgsinos. Gauginos are often called by individual names, since they play different roles in phenomenology. The gaugino partners of the gluon, W, and B are respectively called gluino, wino, and bino. The names, notations for superfield components, and quantum numbers of the MSSM superfields are listed in Tables 2.2 and 2.3.

The superpotential of the MSSM is

$$W_{MSSM} = -y_u^{ij} H_u \circ Q_i \bar{U}_j + y_d^{ij} H_d \circ Q_i \bar{D}_j + y_e^{ij} H_d \circ L_i \bar{E}_j + \mu H_u \circ H_d. \tag{2.82}$$

Table 2.2 The names, superfield components, G_{SM} representation, and $3(B-L)$ assignments of the MSSM chiral supermultiplets

Superfield	Name	Scalar	Spinor	G_{SM} representation	$3(B-L)$
Q	Quark doublet	$\widetilde{q}$	q	$(\mathbf{3}, \mathbf{2}, \frac{1}{6})$	1
$\overline{U}$	Up singlet	$\widetilde{\bar{u}}$	$\bar{u}$	$(\overline{\mathbf{3}}, \mathbf{1}, -\frac{2}{3})$	-1
$\overline{D}$	Down singlet	$\widetilde{\bar{d}}$	$\bar{d}$	$(\overline{\mathbf{3}}, \mathbf{1}, \frac{1}{3})$	-1
L	Lepton doublet	$\widetilde{\ell}$	ℓ	$(\mathbf{1}, \mathbf{2}, -\frac{1}{2})$	-3
$\overline{E}$	Lepton singlet	$\widetilde{\bar{e}}$	$\bar{e}$	$(\mathbf{1}, \mathbf{1}, 1)$	3
H_u	Up-type Higgs	h_u	$\widetilde{h}_u$	$(\mathbf{1}, \mathbf{2}, \frac{1}{2})$	0
H_d	Down-type Higgs	h_d	$\widetilde{h}_d$	$(\mathbf{1}, \mathbf{2}, -\frac{1}{2})$	0

Table 2.3 The names and notations for the superfield components of the MSSM vector supermultiplets

G_{SM} subgroup	Spinor		Vector	
	Name	Symbol	Name	Symbol
U(1)	Bino	$\widetilde{B}$	B boson	B
SU(2)	Wino	$\widetilde{W}$	W boson	W
SU(3)	Gluino	$\widetilde{g}$	Gluon	G

Indices i and j run over three generations of the quark and lepton fields. The Higgs bilinear term $\mu H_u \circ H_d$ has no corresponding interaction in the SM and feeds into the mass of the Higgs particle after EWSB.

The terms in Eq. (2.4) were in fact the full set of renormalizable non-gauge interactions allowed by gauge symmetries in the SM. The situation is different in the MSSM. Besides the Higgs bilinear term, the terms in the B- and L-violating superpotential

$$W_{\not{R}} = \lambda^{ijk} L_i \circ L_j \overline{E}_k + \lambda'^{ijk} L_i \circ Q_j \overline{D}_k + \lambda''^{ijk} \overline{U}_i \overline{D}_j \overline{D}_k + \mu'^{i} L_i \circ H_u \qquad (2.83)$$

are gauge invariant. Such interactions would, for example, cause rapid proton decay and thus need to be severely suppressed or absent.

Suppression of the undesirable terms can be achieved by postulating the conservation of a new parity. This multiplicative quantum number is called matter parity, and is defined by

$$P_{\text{mat}} = (-1)^{3(B-L)}. \qquad (2.84)$$

All of the terms in Eq. (2.82) have $P_{\text{mat}} = 1$, while the B- and L-violating terms have $P_{\text{mat}} = -1$, and therefore should be absent.

As a useful mnemonic quantity, another parity named R-parity

$$P_R = (-1)^{3(B-L)+2s} \qquad (2.85)$$

is defined for individual components of the MSSM superfields, with the spin s of the fields. R-parity is physically exactly equivalent to matter parity, since the fermions always appear in pairs in the Lagrangian and therefore the sum of $2s$ for any term must be even. Its usefulness is given by the R-parity of sfermions, higgsinos, and gauginos being odd, while all the SM fields and h_d are R-even. Therefore R-parity makes it clear that non-SM particles always appear in even numbers in any interaction if matter parity is conserved.

An important consequence of the matter parity conservation is that the R-odd particles, collectively referred to as sparticles, are produced and annihilated at least in pairs. Furthermore, a sparticle can only decay into another sparticle and an R-even combination of particles. In particular, this implies that the lightest sparticle (LSP) is stable. If sparticles were produced copiously during the Big Bang, the LSP must be abundant in the universe. The fact that no LSP candidate has been observed so far suggests that it is very weakly interacting. In other words, the MSSM naturally provides a candidate for the dark matter as a by-product.

A complete Lagrangian of the MSSM must include the soft SUSY-breaking terms. Possible terms respecting gauge symmetries and matter parity are

$$
\begin{aligned}
\mathcal{L}_{\text{soft}} = &-\frac{1}{2}\left(M_3 \widetilde{g}\widetilde{g} + M_2 \widetilde{W}\widetilde{W} + M_1 \widetilde{B}\widetilde{B} + \text{c.c.}\right) \\
&-\left(-a_u^{ij} h_u \circ \widetilde{q}_i \widetilde{u}_j + a_d^{ij} h_d \circ \widetilde{q}_i \widetilde{d}_j + a_e^{ij} h_d \circ \widetilde{\ell}_i \widetilde{e}_j + \text{c.c.}\right) \\
&-\left((m_q^2)^{ij} \widetilde{q}_i^* \widetilde{q}_j + (m_\ell^2)^{ij} \widetilde{\ell}_i^* \widetilde{\ell}_j + (m_u^2)^{ij} \widetilde{u}_i^* \widetilde{u}_j + (m_d^2)^{ij} \widetilde{d}_i^* \widetilde{d}_j + (m_e^2)^{ij} \widetilde{e}_i^* \widetilde{e}_j\right) \\
&-\left(m_{h_u}^2 h_u^* h_u + m_{h_d}^2 h_d^* h_d\right) - \left(b h_u h_d + \text{c.c.}\right).
\end{aligned}
\tag{2.86}
$$

Equation (2.86) includes mass terms for gauginos and sfermions, but not for SM fermions. Thus the feature of the SM that the masses of the fermions are generated through EWSB is still present in the MSSM. On the other hand, gauginos and sfermions have tree-level masses given in Eq. (2.86). The higgsinos also have a tree-level mass of $|\mu|^2$. Since these masses can be heavier than the Z boson mass m_Z, which is the characteristic scale of EWSB, it may be natural that the sparticles have not been observed so far.

In fact, taking the opposite approach on the same point, requiring that the sparticle masses be heavy actually necessitates the mechanism of indirect SUSY breaking discussed at the end of Sect. 2.7. The reason originates from a sum rule for squared masses of all the particles directly coupled to each other within a given supersymmetric Lagrangian. The sum rule applies generally to mass spectra of exact or spontaneously broken SUSY, and dictates that the lightest scalar in the theory must be lighter than any fermions. This result can be avoided if the SUSY-breaking order parameter belongs to a Lagrangian that does not have direct interactions with MSSM fields, i.e., if the SUSY breaking is communicated only through some intermediate interactions to the MSSM.

2.9 Soft Supersymmetry-Breaking Parameters

As already seen in Sect. 2.7, under the assumption that the terms in $\mathcal{L}_{\text{soft}}$ originate from a single SUSY-breaking dynamics and a common messenger interaction,

$$M_3, M_2, M_1, a_u, a_d, a_e \sim M_{\text{soft}} \tag{2.87}$$

and

$$m_{\tilde{q}}^2, m_{\tilde{\ell}}^2, m_{\tilde{u}}^2, m_{\tilde{d}}^2, m_{\tilde{e}}^2, m_{h_u}^2, m_{h_d}^2, b \sim M_{\text{soft}}^2 \tag{2.88}$$

should hold. Since $m_{h_u}^2$, $m_{h_d}^2$, and b together with μ drive the EWSB, M_{soft} cannot be too much higher than $\mathcal{O}(1)$ TeV. This is the baseline argument for expecting to find sparticles in particle colliders such as the CERN Large Hadron Collider.

Interestingly, this postulate of the so-called TeV-scale SUSY is a critical ingredient of an unrelated prediction of the MSSM, the unification of gauge couplings. The running of the gauge couplings g_1, g_2, and g_3 differs between the MSSM and the SM, because the former has more charged degrees of freedom than the latter. Since the rate of running is determined by the particles in the loops of the vacuum polarization calculations, to a good approximation the scale M_{soft} serves as the threshold at which this rate switches from that in the SM to the MSSM. The evolutions of the coupling constants traced up from weak-scale values under this framework with three different M_{soft} are shown in Fig. 2.2. Here, the RG equations are at one-loop order, given by

$$\frac{d}{d(\ln\frac{Q}{Q_0})}\alpha_A^{-1} = -\frac{b_A}{2\pi}, \quad (b_1, b_2, b_3) = \begin{cases} (\frac{41}{10}, -\frac{19}{6}, -7) & Q < M_{\text{soft}} \\ (\frac{33}{5}, 1, -3) & Q > M_{\text{soft}} \end{cases} \tag{2.89}$$

In Eq. (2.89), Q and Q_0 are renormalization and reference scales, respectively. From the graphs, it is apparent that the three couplings meet only when $M_{\text{soft}} \lesssim 10^{3-4}$ GeV. Such an observation is obviously at best a circumstantial evidence, but is nevertheless an encouragement, for TeV-scale SUSY. The fact that the three couplings meet, paving a way to the grand unification of the forces is sometimes considered as a motivation for the MSSM which is as strong as the suppression of the quadratic divergence of the Higgs mass.

Once the electroweak symmetry is broken, the electric charge is the only remaining quantum number of the wino, the bino, and the higgsino fields. The wino and higgsino charge eigenfields in the unitary gauge are

$$\widetilde{W}^{\pm} := \frac{1}{\sqrt{2}}(\widetilde{W}^1 \mp \widetilde{W}^2), \quad \widetilde{W}^0 := \widetilde{W}^3, \quad h_u = (h_u{}^+, h_u{}^0), \quad h_d = \begin{pmatrix} h_d{}^0 \\ h_d{}^- \end{pmatrix}. \tag{2.90}$$

Fig. 2.2 One-loop renormalization group evolutions of $\alpha_A^{-1} = 4\pi/g_A^2$ as functions of the renormalization scale Q. The input values at m_Z are $\alpha_1^{-1} = 59.2$, $\alpha_2^{-1} = 29.6$, and $\alpha_3^{-1} = 8.55$. The inset shows a close-up view of 10^{15} GeV $< Q < 10^{17}$ GeV

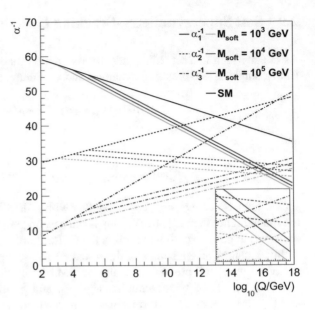

The gaugino-fermion bilinear terms in $\mathcal{L}_{\mathrm{SuperYM}}$ cause these fields to mix and form mass eigenfields. The resulting four neutral and four charged fermions are referred to as the neutralinos and the charginos, respectively. The neutralinos and the charginos are collectively called the electroweakinos. In the basis $\Psi^0 = (\widetilde{B}, \widetilde{W}^0, \widetilde{h}_d^{\,0}, \widetilde{h}_u^{\,0})$, the neutralino mass terms are

$$\mathcal{L}_{\mathrm{neut.}} = -\frac{1}{2}(\Psi^0)^T \mathbf{M_N} \Psi^0 + \text{c.c.}, \tag{2.91}$$

where the mixing matrix $\mathbf{M_N}$ is

$$\mathbf{M_N} = \begin{pmatrix} M_1 & 0 & -g_1 v_d/2 & g_1 v_u/2 \\ 0 & M_2 & g_2 v_d/2 & -g_2 v_u/2 \\ -g_1 v_d/2 & g_2 v_d/2 & 0 & -\mu \\ g_1 v_u/2 & -g_2 v_u/2 & -\mu & 0 \end{pmatrix}, \tag{2.92}$$

with the Higgs VEV $v_u := \sqrt{2}\langle h_u \rangle$ and $v_d := \sqrt{2}\langle h_d \rangle$. Similarly, chargino mass terms in the basis $\Psi^{\pm} = (\widetilde{W}^{\pm}, \widetilde{h}_{[ud]}^{\pm})$ are

$$\mathcal{L}_{\mathrm{chrg.}} = -\frac{1}{2}(\Psi^+)^T \mathbf{M_C} \Psi^- + \text{c.c.}, \tag{2.93}$$

with the mixing matrix

$$\mathbf{M_C} = \begin{pmatrix} M_2 & g_2 v_d/\sqrt{2} \\ g_2 v_u/\sqrt{2} & \mu \end{pmatrix}. \tag{2.94}$$

The forms of the two matrices indicate that the mixing between gauginos and higgsinos is controlled by the Higgs VEV. If $M_1, M_2, |\mu| \gg v_u, v_d$, then neutralinos and charginos are approximately purely wino, bino, and higgsinos. Within such a hierarchy, depending on which of M_1, M_2, and $|\mu|$ is the smallest, lightest neutralino $\widetilde{\chi}_1^0$ and chargino $\widetilde{\chi}^{\pm}_1$ are said to be bino-like, wino-like, or higgsino-like, respectively. The identities of the lightest electroweakinos are highly important for phenomenology.

Another point to note about the soft SUSY-breaking Lagrangian is that a large fraction of its more than one hundred free parameters must be severely constrained to avoid enhancing various flavor-changing and CP-violating interactions beyond observed limits. In principle, the sfermion masses and a-parameters can have arbitrary off-diagonal components in the sfermion family basis and can generate flavor-changing neutral currents. Additionally, complex phases in the a-parameters can be a source of CP violation.[2] Therefore, some explanation is called for to reconcile $\mathcal{L}_{\text{soft}}$ with the known rareness of flavor and CP violations.

It is possible that the sfermions are extremely heavy. In this case, flavor-violating effects are automatically suppressed since they arise from loops. However, this scenario goes against the assumption from Eqs. (2.87) and (2.88) and might suggest that either $M_{\text{soft}} \gg m_Z$ or that an additional SUSY-breaking effect exists that couples only to sfermions.

The more commonly held assumption is that the SUSY-breaking effect respects the flavor structure of the SM. In practice this means that the sfermion masses are degenerate, given by

$$
(m_q^2)^{ij} = M_q^2 \delta^{ij}, \ (m_\ell^2)^{ij} = M_\ell^2 \delta^{ij},
$$
$$
(m_u^2)^{ij} = M_u^2 \delta^{ij}, \ (m_d^2)^{ij} = M_d^2 \delta^{ij}, \ (m_e^2)^{ij} = M_e^2 \delta^{ij},
$$
$$\tag{2.95}$$

and the a-terms are proportional to the Yukawa couplings in the superpotential:

$$
a_u^{ij} = A_u y_u^{ij}, \ a_d^{ij} = A_d y_d^{ij}, \ a_e^{ij} = A_e y_e^{ij}. \tag{2.96}
$$

These conditions, however, should be regarded as boundary conditions at some input scale for the RG evolution of the soft parameters. The reason is as follows. Whatever the relations that the soft parameters satisfy at the input scale, they will be somewhat modified by the RG-running down to M_{soft}. Therefore, demanding that Eqs. (2.95) and (2.96) hold at the output scale implies a fine-tuning of the SUSY-breaking and/or communication dynamics. It is far more natural to assume that the SUSY-breaking sector and the messenger interaction are suitably flavor-conserving, resulting in the conditions above as inputs to the MSSM. Conversely, even with flavor-conserving input, the MSSM gives predictions on, e.g., rates of flavor-violating decays of heavy leptons that can be largely different from that in the SM [10, 11].

---

[2]It is assumed that the sfermion generations are defined by the partner fermion mass eigenfields, and the complex phases of the fields are defined so that gaugino masses and b are real.

The input scale is defined by the mechanism for the communication of SUSY breaking. Then, in regard to the grand unification discussed above, messenger interactions that set input scales above $M_{\text{GUT}} \sim 10^{16}$ GeV can be problematic. The reason is that at such input scales there would be no distinction between squarks and sleptons, resulting in fewer free parameters than there are in Eqs. (2.95) and (2.96). Consequently, as the parameters are RG-evolved down below M_{GUT}, uncontrollable mixings can invalidate these relations.

The requirements in Eqs. (2.95) and (2.96), driven by the non-observation of large flavor-changing effects, are the strongest clue for the nature of the SUSY-breaking and its communication to the MSSM. The next section discusses one of the general communication frameworks that can satisfy these requirements naturally.

2.10   Gauge-Mediated Supersymmetry Breaking

Gauge-mediated SUSY breaking (GMSB) is one of the oldest and most well-studied mechanisms for generating $\mathcal{L}_{\text{soft}}$ of the MSSM. The basic idea is to conceive a "hidden sector" which has no direct interaction with the MSSM fields, and demand that SUSY is spontaneously broken by some VEV in this sector. Additionally, either this SUSY-breaking field or another set of fields that couple to it must be charged under G_{SM}. The gauge interactions then act as the indirect communication channel between the hidden sector and the MSSM. The MSSM soft SUSY-breaking terms emerge as a result of integrating out the hidden degrees of freedom.

The virtue of this framework is that Eq. (2.95) is automatically satisfied because the gauge interactions do not discriminate quark and lepton generations. Furthermore, unlike Planck-scale SUSY breaking, which is another well-studied mechanism, the input scale where this flavor-blindness condition is applied can be lower than the GUT scale, circumventing the problem of GUT-originated flavor mixing discussed in Sect. 2.9. An input scale below M_{GUT} is also advantageous because a-parameters are in general small at the input in GMSB, and their weak-scale values are dominantly due to inhomogeneous RG evolution, which will be proportional to the Yukawa couplings. Therefore, the condition in Eq. (2.96) is also naturally fulfilled.

In the minimal implementation of GMSB, the SUSY-breaking field S is a singlet under G_{SM}. It couples to the messenger sector, or a set of charged fields $\{\Phi_I, \overline{\Phi}_I\}$, via a superpotential

$$W_{\text{mess.}} = \sum_I y_I S \Phi_I \overline{\Phi}_I. \tag{2.97}$$

For each I, Φ_I and $\overline{\Phi}_I$ are in mutually conjugate representations of G_{SM} (for U(1), they have opposite hypercharges). When the scalar component of S develops a VEV, the mass degeneracy of the Φ_I and $\overline{\Phi}_I$ supermultiplets is also resolved. Then, the scalars and fermions of Φ_I in turn contribute non-canceling radiative corrections to the gaugino two-point function, giving gauginos their mass. The same radiative corrections are applied to two-point functions of the scalars of the chiral multiplets. The a-parameters arise from higher-order corrections and can be ignored. The value of the b-parameter depends on more than the simple parametrized dynamics of Eq. (2.97) and is not part of the minimal GMSB.

In minimal GMSB, the ratios of the gaugino masses are fixed by the gauge coupling constants:

$$M_A = \frac{\alpha_A}{4\pi}\Lambda,\tag{2.98}$$

where $A = 1, 2, 3$ and the common parameter Λ encodes the effect of integrating out the SUSY-breaking and messenger sectors. The scalar soft masses are also proportional to Λ, with coefficients containing group-theoretic constants that depend on the representation of G_{SM} the fields belong to. In short, the soft masses in the minimal GMSB are completely determined by a single parameter. Such a model would be highly predictive, but perhaps not very realistic.

A less restrictive, general formulation of gauge mediation is given in [12]. In this framework, dubbed general gauge mediation (GGM), a model is considered to be based on GMSB if, in the limit of $\alpha_A \to 0$, it decouples to an unbroken MSSM and another sector which contains SUSY breaking. The authors analyzed the most general form of such models by representing the hidden sector with a conserved current. As a result, it was shown that the full GMSB model space is spanned by a set of 3 real and 1 complex parameter per gauge subgroup plus a mass scale. In terms of the MSSM soft terms, to lowest order in α_A, the 9 real parameters combine into 3 constants and define the sfermion masses, which are still generation-blind, and the complex parameters give the gaugino mass for each subgroup. Since there are no a priori relations implied between the parameters, in particular the spectrum of the gaugino masses is not fixed by the gauge coupling strengths, which was the case in Eq. (2.98).

Even in GGM, the RG evolution of the mass parameter, which is more rapid for colored particles, tends to push squarks heavier than the sleptons when evaluated at the weak scale. Similarly, the gluino tends to be the heaviest of the three gauginos regardless of the precise relations at the input scale. Within the sfermions, the mass of individual generations is driven by their Yukawa couplings and the a-parameters. The mass degeneracy between the generations is resolved through this RG evolution, and commonly results in a spectrum where the third-generation sfermions are lighter than their first- and second-generation counterparts.

It should be noted that the general mass relations above are applicable to any SUSY-breaking model. When considering the LSP, however, GMSB shows a peculiar feature that is not commonly realized in other breaking mechanisms. In GMSB, the gravitino, the spin-$3/2$ component of the graviton superfield, becomes much lighter than any of the MSSM particles.

To better understand this statement, local supersymmetry must be considered, in which the spinor parameter ζ in Eq. (2.30) is a function on spacetime. In other words, ζ is made into a gauge degree of freedom. A locally supersymmetric Lagrangian must contain a spin-3/2 field $\widetilde{\mathcal{G}}$ that transforms inhomogenously under SUSY, as

$$\delta_\zeta \widetilde{\mathcal{G}}_\mu^a = \partial_\mu \zeta^a + \dots \tag{2.99}$$

This field couples to all superfields, i.e., to all sources of energy-momentum. Given that SUSY itself is a spacetime symmetry, the only conclusion is that $\widetilde{\mathcal{G}}$ must be the gravitino, the supersymmetric partner of the graviton $\mathcal{G}$. The gravitino is massless in unbroken SUSY, but it acquires mass through the VEV of a SUSY-breaking auxiliary field. From a general argument, the gravitino mass $m_{3/2}$ will scale as

$$m_{3/2} \sim \frac{\langle F \rangle}{M_{\text{Pl}}} \tag{2.100}$$

because it should be zero either when SUSY is not broken or when gravity is turned off. On the other hand, Eq. (2.81) relates the scale of the soft mass parameters of the MSSM to $\langle F \rangle$. Therefore, Eq. (2.100) can also be written as

$$m_{3/2} \sim M_{\text{soft}} \frac{M_{\text{mess.}}}{M_{\text{Pl}}}. \tag{2.101}$$

With $M_{\text{soft}} \lesssim \mathcal{O}(1)$ TeV and $M_{\text{mess.}} < M_{\text{GUT}}$,

$$m_{3/2} \lesssim \mathcal{O}(1) \, \text{GeV}. \tag{2.102}$$

The condition (2.102) is a special feature of GMSB; in e.g., Planck-scale mediation, $M_{\text{mess.}} = M_{\text{Pl}}$ and therefore $m_{3/2} \sim M_{\text{soft}}$. Since the gravitino is R-odd and no other R-odd particles are expected to exist below the weak scale, the gravitino in GMSB is the LSP.

When SUSY is spontaneously broken, the fermion field that accompanies the auxiliary field that acquires a VEV becomes a massless degree of freedom called the goldstino. In local SUSY, the "gauge" can be fixed so that the goldstino becomes the helicity $\pm 1/2$ components of the gravitino. The interaction strengths of the goldstino with the other fields scale as

$$g_{\text{MSSM-}\widetilde{\mathcal{G}}} \sim \frac{\Delta m^2}{\langle F \rangle}, \tag{2.103}$$

where Δm^2 is the mass splittings between the superpartners. The inverse proportionality to $\langle F \rangle$ can be understood from Eq. (2.44), which indicates that the on-shell field amplitude of the goldstino is proportional to $\langle F \rangle$. The magnitude of the goldstino interaction terms is then one power lower in $\langle F \rangle$ than the free goldstino Lagrangian, and therefore the interaction strength is suppressed by that much. The factor in the numerator regulates the divergence when $\langle F \rangle \to 0$, since the mass splittings also vanish at that limit. Alternatively, using $\Delta m^2 \sim M_{\text{soft}}^2$ and Eq. (2.81) gives

$$g_{\text{MSSM-}\widetilde{g}} \sim \frac{M_{\text{soft}}}{M_{\text{mess.}}}. \tag{2.104}$$

It has been pointed out [13] that the Higgs mass of 126 GeV implies a rather large weak-scale value for a_t, the a-parameter for the superpartner of top, or stop. Since a-parameters are expected to be negligibly small at the input scale, it follows that there is a large gap between the messenger scale and the weak scale to generate the required value through RG evolution. Additionally, the gluino must be heavy since its mass is a major driving factor for the evolution of a_t. According to the numerical analysis performed by the authors, a multi-TeV gluino is favored to keep the messenger scale below M_{GUT}. From Eqs. (2.101) and (2.104), it can be concluded that currently a weakly interacting, heavy goldstino–gravitino is favored. The phenomenological implications of this point will be discussed in Sect. 2.12.

2.11 The Phenomenology of the MSSM

As a model of nature, TeV-scale SUSY must make verifiable predictions which can be tested by observations. The experimental and observational tests of SUSY can be, roughly speaking, categorized into five types: Those where some observable is measured to find possible deviation from the SM prediction; those where the scattering of the dark matter particles with atomic nuclei is explored; dark matter annihilation searches; cosmological observations; and those where the direct production of sparticles is involved.

The first type probes radiative corrections from new particles, and includes the measurement of the rates of rare decays already mentioned in Sect. 2.9. In particular, the decay rate of $\mu \rightarrow e\gamma$ is expected to be practically unmeasurable in the SM. Thus an observation of such a decay implies higher-order corrections from physics beyond the SM, which can be attributed to nondiagonal mass matrices of the sleptons. Another decay rate measurement is motivated by the existence of two Higgs fields in the MSSM. References [14, 15] showed that effective operators in the two-Higgs doublet model, including the MSSM, enhance the decays of neutral B mesons, or mesons containing one (anti-)b quark, to lepton pairs. Specifically, the decay $B_s^0 \rightarrow \mu^+\mu^-$ has been a subject of heavy scrutiny, since its rate in the SM is calculable with fair precision and is sufficiently high to be measurable. The tests of the first type that are not decay rate measurements include the measurement of the anomalous magnetic moment of the muon (a_μ), which also gets higher-order MSSM contributions owing to the existence of two Higgs fields.

The tests of the second, third, and fourth type are focused on the prediction of the LSP dark matter in R-parity conserving MSSM. They are therefore not tests of MSSM per se, but their results give important constraints on the MSSM. The dark matter scattering experiments probe the mass of the dark matter particles and their interaction strength with the normal matter directly and largely model-independently. The annihilation search seeks for signs of dark matter interacting with itself, producing SM particles [16]. It probes the mass and the annihilation cross section of the dark matter. The cosmological observations of, e.g., the cosmic microwave background, on the other hand, require a model of the early universe for them to be translated into conclusions on dark matter physics. Nevertheless, within specific frameworks, they are also capable of constraining the mass and annihilation cross section of dark matter from thermodynamical arguments.

A general restriction on GMSB, in particular its gravitino LSP, from cosmology is set in two ways. On one hand, if the gravitino is too light, it becomes the relativistic dark matter which wipes out the structure formation in the early universe [17]. On the other hand, an over-abundance of the heavy gravitino would cause the dark matter relic density to exceed the value derived from cosmological observations. Thus a gravitino of mass above $\mathcal{O}(10^{-4})$ GeV implies an upper bound on the reheating temperature to not overproduce sparticles, and with them the gravitino [18, 19].

The direct-production experiments, which is the last type of experiments listed above, are performed using high-energy particle colliders. When interacting particles are brought to a very small volume in space with sufficiently high energy, particles different from the incoming ones can emerge with probabilities governed by the Lagrangian of nature. In high-energy colliders, two beams of stable particles in opposite directions are concentrated at an interaction point, which is surrounded by detectors to measure and identify the outgoing particles. A collider is characterized by the type of beam particles and the center-of-mass energy ($\sqrt{s}$). The notation $\sqrt{s}$ derives from the Mandelstam variable $s = (k_1 + k_2)^2$, where k_1 and k_2 are the 4-momenta of the particles in the two-body physics process. So far there have been three different types of particle colliders relevant for SUSY searches: the electron–positron (e^-e^+) collider, which can study production of sparticle pairs that are electrically neutral; the proton–electron (pe) collider, which can study processes initiated by electron–quark interactions; and the hadron collider, which can be a proton–proton (pp) or a proton–antiproton ($p\bar{p}$) machine, and probes the interactions of quarks and gluons.

The cross sections for the production processes of most of the sparticles are calculable with good accuracy as functions of the sparticle masses, as demonstrated in Fig. 2.3, since most of the unknown parameters of the MSSM are related to sparticle masses and not to interaction strengths. This property can be used to distinguish the MSSM from other models as an explanation to newly found particles, should there be any. Once a heavy sparticle is produced, assuming R-parity conservation, it will decay into a multitude of stable SM particles plus an odd number of LSPs, unless the produced sparticle is the LSP itself. The expected heaviness of the sparticle usually demands that the decay is prompt, i.e., the full decay cascade takes place within the microscopic region at the interaction point. However, it is worth mentioning that there are specific models that predict non-prompt decays. In such models, "displaced" signals are observed, where certain particles emerge from a point separated from the interaction point by a macroscopic distance, or even more spectacularly, metastable sparticles directly arrive at the detector.

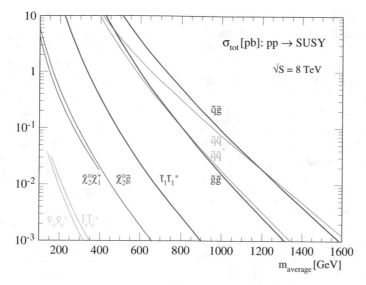

Fig. 2.3 Cross sections for producing various pairs of sparticles at a proton–proton collider with $\sqrt{s} = 8\,\mathrm{TeV}$, calculated at next-to-leading order in QCD with the `Prospino2` calculator [20]. (Source: http://www.thphys.uni-heidelberg.de/~plehn/)

Limiting the discussion to promptly decaying sparticles for now, the production of a sparticle is inferred from the multiplicity, momenta, and species of the decay products. Such "signatures" can be predicted for each mass spectrum of the MSSM; the central concern of collider phenomenology is to predict the distinctive signatures of the models of interest. From an experimental perspective, then, the most urgent task is to detect the outgoing stable decay products as completely as possible and reconstruct the primary process that took place at the interaction point. In modern colliders, nearly hermetic particle detectors are built around the interaction point, with only the region around the incoming beams lacking sensitivity. The detectors are typically multi-layered to exploit different technologies for detecting different particles. An important limitation here is that all of the technologies must ultimately rely on either electromagnetic or strong forces. This limitation in particular implies that the LSP would not be detected. Therefore, missing momentum, i.e., a significant deficit in the detected outgoing momentum compared to what is expected from the input, is a quintessential signal of sparticle production. At lepton colliders, the center-of-momentum system of the interaction is at rest, and the missing momentum can be truly calculated as a missing 4-vector. In contrast, the primary interaction at the hadron colliders is between quarks and gluons that constitute the hadrons, which makes it impossible to know the momentum of the interaction system along the beam (longitudinal) direction. Accordingly, the imbalance of the observed final-state momenta in the transverse direction, called missing transverse momentum or missing transverse energy ($E_\mathrm{T}^\mathrm{miss}$), is the closest substitute to the full missing momentum.

The collider signature in the GMSB scenario is primarily defined by the next-to-lightest sparticle (NLSP). This general feature is a direct consequence of Eq. (2.104) and R-parity conservation. The $1/\langle F \rangle$-suppression of the MSSM-goldstino coupling makes the interaction of the goldstino–gravitino to other particles significantly weaker than intra-MSSM interactions.[3] Therefore all sparticles decay to another sparticle that is not the goldstino, except for the NLSP whose only kinematically allowed and R-parity conserving decay is to its SM partner and the goldstino. GMSB sparticle production will, then, always feature the SM partner of the NLSP in the final state along with the missing momentum due to the goldstino–gravitino. An exception is when $\langle F \rangle$ is very large, such that the MSSM-goldstino coupling is suppressed to the point that the NLSP is long-lived or collider-stable. In this case the NLSP will feature displaced decays, or if it is neutral and collider-stable, simply appear as missing momentum.

In the minimal GMSB, Eq. (2.98) dictates that if the NLSP is a gaugino, it would be a bino. The smoking gun for sparticle production in the bino-NLSP scenario is the signature of two photons and large missing momentum, which is easily identifiable and is fairly rare in the SM. If the sleptons are instead lighter than the gauginos, the right-handed stau, which is the scalar particle of the third-generation $\overline{E}$, would likely assume the role of the NLSP. Right-handed stau is the primary NLSP candidate in this case since SU(2) and SU(3) interactions tend to push the mass up in the RG evolution, while the Yukawa interactions do the opposite. The signature in the stau-NLSP scenario would be two τ leptons and missing momentum.

In the GGM framework, Eq. (2.98) does not need to hold anymore, and any sparticle can in principle be the NLSP. It would still be true from RG arguments that colored particles would be heavier than non-colored ones, but there is no reason that, e.g., the wino cannot be the NLSP. Assuming prompt decays, GGM therefore calls for a systematic check of all supposable $\widetilde{X} \to X + \widetilde{\mathcal{G}}$ patterns, where $\widetilde{X}$ and X are the NLSP and its SM partner.

Of particular interest are the general neutralino-NLSP scenarios [21]. As shown in Sect. 2.9, a neutralino can be bino-like, wino-like, higgsino-like, or a more general superposition of the three. For phenomenological discussions, it is easier to focus on the first three cases. Observables of the last case can be recovered by weighted mixtures of those from the first three. The bino-NLSP case is similar to the minimal GMSB. For the higgsino NLSP, it is important to note that the NLSP is in general a superposition of $\widetilde{h}_u$ and $\widetilde{h}_d$, with the mixing angle depending on the details of the MSSM parameters related to EWSB. The preferred decay mode of the NLSP also varies with these parameters. If the NLSP has a high fraction of the superpartner of the neutral EWSB Nambu–Goldstone boson[4] appearing in Eqs. (2.13) and (2.14), it decays strongly to (the longitudinal component of) Z and $\widetilde{\mathcal{G}}$. If instead the NLSP

---

[3]The goldstino–gravitino also exhibits gravitational interactions with the other particles, but it would be even weaker than the goldstino interaction.

[4]The superpartner of the EWSB Nambu–Goldstone boson is unrelated to the goldstino, which is the superpartner of some SUSY-breaking scalar field.

Fig. 2.4 BR of a pure-wino neutralino NLSP as a function of the NLSP mass. Reprinted from [21]

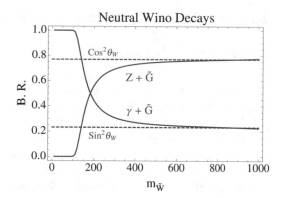

has a larger component of the superpartner of the Higgs boson, the decay mode $h\widetilde{\mathcal{G}}$ is preferred. The collider signature of the latter case is two sets of Higgs decay products plus missing momentum. The rich variety of Higgs decay modes presents many options to choose the search signature from.

If the neutral wino is the NLSP, it is likely that charged winos are as light as the NLSP. In this so-called wino co-NLSP scenario, the winos are produced in a collider in either of the combinations $\widetilde{\chi}^{\pm}\widetilde{\chi}^{\pm}$, $\widetilde{\chi}^0\widetilde{\chi}^0$, or $\widetilde{\chi}^{\pm}\widetilde{\chi}^0$. The last combination has a net charge and thus is available at lepton colliders only as the next-to-final stage of some SUSY decay cascade. A charged wino NLSP always decays to $W^{\pm}\widetilde{\mathcal{G}}$, whereas its neutral counterpart can go to either $Z\widetilde{\mathcal{G}}$ or $\gamma\widetilde{\mathcal{G}}$. The decay BR of pure-wino neutralino was calculated in [21] as a function of the wino mass, and is shown in Fig. 2.4. In the high-mass limit where the kinematic factor of m_Z can be ignored, the decays are simple projections of W'^3 in Eqs. (2.17) and (2.22) to its A and Z components, and thus the BR reaches $\sin^2\theta_W$ and $\cos^2\theta_W$, respectively. The observable signature in the wino co-NLSP scenario is limited. Particularly, at hadron colliders, events where the electroweak bosons W or Z decay hadronically would be very difficult to distinguish from the overwhelming background of vector boson plus QCD hadrons. Therefore, a search of, e.g., $\widetilde{\chi}^{\pm}\widetilde{\chi}^0$ would be conducted in the channels $\widetilde{\chi}^{\pm}\widetilde{\chi}^0 \rightarrow W^{\pm}Z\widetilde{\mathcal{G}}\widetilde{\mathcal{G}} \rightarrow \ell^{\pm}\nu\ell^{\pm}\ell^{\mp}\widetilde{\mathcal{G}}\widetilde{\mathcal{G}}$ (multi-lepton) or $\widetilde{\chi}^{\pm}\widetilde{\chi}^0 \rightarrow W^{\pm}\gamma\widetilde{\mathcal{G}}\widetilde{\mathcal{G}} \rightarrow \ell^{\pm}\nu\gamma\widetilde{\mathcal{G}}\widetilde{\mathcal{G}}$ (photon–lepton). Since the BR of the Z to two observable leptons e and μ is 6% in total, it follows that the latter has a higher acceptance than the former.

Before concluding this section, it should be mentioned that the Higgs mass of 126 GeV puts a strong constraint on the MSSM [22], including the GGM scenarios. Draper et al. [13] showed that the observed mass of the Higgs, generally considered somewhat high for the MSSM, indeed requires a large A_t value at the electroweak scale. For GMSB, where $A_t \ll \sqrt{m_{\tilde{u}}^2}$ at the input scale, this in turn implies a high messenger scale to generate sufficiently large A_t through the RG evolution. Then, to keep M_{soft} not too much higher than $\mathcal{O}(1)$ TeV, $\langle F \rangle$ must also be high. Numerical calculation by the authors revealed that unless the gluino mass is in multi-TeV range, a 126 GeV Higgs suggests a heavy, very weakly interacting gravitino for GMSB. Thus, according to the article, the GMSB signal is either non-prompt, or prompt but does not involve gluino production.

2.12   The Status of Searches for Supersymmetry

2.12.1   General Status

At present, there is no evidence for a supersymmetric extension of the standard model. As seen in Sect. 2.11, TeV-scale SUSY would leave its "footprints" in multiple observable phenomena. Various direct and indirect searches are complementary, probing different aspects of the MSSM and narrowing down the parameter space from multiple fronts. However, without inputs on some critical parameters, it is also fair to say that very few definitive statements can be made.

Even if a precise picture is difficult to draw, some general trend can be stated from the existing results, keeping in mind that it can always be overturned in specific regions of the parameter space. For example, the $\mu \to e\gamma$ process has not been observed yet, with the upper limit on the decay BR at 5.7×10^{-13} [23]. The decay $B_s^0 \to \mu^+\mu^-$ has recently been observed, but the BR $\left(2.8^{+0.7}_{-0.6}\right) \times 10^{-9}$ [24] is consistent with the SM calculation $(3.66 \pm 0.23) \times 10^{-9}$ within errors [24]. These two results point in general to heavier sparticles or smaller $\tan \beta := v_u/v_d$, since the rates of these decays are enhanced when the effective coupling of H_u with down-type $(T_3 = -1/2)$ fermions is large and the sparticles involved in the radiative corrections are light.

A result somewhat in tension with the above is the measurement of a_μ. The current best measurement of a_μ is in excess of the SM prediction by 2σ [25]. If this discrepancy is to be solely explained by SUSY, there would be a positive relation between $\tan \beta$ and the upper bound on the slepton mass. As null results from direct searches push lower bound on the slepton mass higher, the required $\tan \beta$ value might exceed the expectation from the rare decay measurements above.

The situation regarding searches for dark matter scattering and annihilation is, at the moment, unclear. Some experiments report possible observations of dark matter scattering [26–28] while others exclude the mass and cross section regions where such signals would occur. In general, increasing sensitivity of direct searches tends to exclude weakly interacting dark matter with mass comparable to those of heavy nuclei most strongly, unless the reported scattering events are indeed due to the dark matter. On the annihilation side, spaceborne experiments [29–31] have probed the excess of positron, antiproton, and gamma ray flux as potential signals of dark matter interactions. The signal must not be pointing to specific astronomical objects, and should feature a cutoff in the energy spectrum, with the endpoint corresponding to the dark matter mass. The Alpha Magnetic Spectrometer on the International Space Station experiment reported a hint of such a cutoff structure at 400 GeV in the positron energy distribution [31], but no definitive conclusions have been drawn yet.

The strongest constraints for TeV-scale SUSY arguably come from collider experiments. SUSY searches have been a high-profile physics program since the late 1980s.

The Large Electron–Positron Collider (LEP) at the European Organization for Nuclear Research (CERN) was an e^-e^+ collider with $\sqrt{s}$ between 90 and 209 GeV, operating from 1989 until 2000. From the result of the highest-energy runs, its four detectors ALEPH, DELPHI, OPAL, and L3 set lower limits on the mass of charginos, sleptons, and third-generation squarks of around 100 GeV [32]. These results are semi-generic and hold unless the sparticle mass spectrum is particularly compressed or some sparticle decays are non-prompt.

The proton–electron collider Hadron-Elektron-Ringanlage (HERA) at the Deutsches Elektronen-Synchrotron (DESY) laboratory in Hamburg operated from 1994 to 2007, with two of its experiments H1 and ZEUS participating in searches for SUSY. Since an electron interacts only with quarks inside a proton, and the possible interaction processes are limited, data from HERA can probe a unique domain of the MSSM, namely squark-slepton production and R-parity violating electron–quark annihilation. The latter interaction arises from the term $\lambda'^{1jk} L_1 \circ Q_j \overline{D}_k$ in Eq. (2.83), with the lepton index fixed to 1 for the incoming e. HERA collided both pe$^+$ and pe$^-$, and from each type of run the model-dependent upper limits were set on $|\lambda'^{1j1}|$ and $|\lambda'^{11k}|$, respectively [33].

Some LEP limits were superseded later by results from the Tevatron, built at Fermilab in Batavia, Illinois, which collided protons with antiprotons at $\sqrt{s}$ of 1.6, 1.8, and 1.96 TeV. Being a p$\overline{p}$ machine, SUSY production at the Tevatron would occur strongly in q$\overline{q}$ interactions at the leading order. Consequently, the two Tevatron experiments CDF and D0 set strong limits on the squark and gaugino productions [34]. The lower limit for the third-generation squark mass was pushed up to around 240 GeV, and the chargino to 170 GeV under specific assumptions.

These situations largely changed with the advent of the Large Hadron Collider (LHC), a pp collider which replaced LEP and started physics operation in 2009. During its Run I, which took place between 2010 and 2012, the LHC ran at $\sqrt{s}$ of 7 and 8 TeV, delivering collision data at an unprecedented interaction rate to each of the two general-purpose detectors ATLAS and CMS. With large data sets of highest-energy particle collisions ever achieved, the two experiments rapidly excluded the regions in the SUSY parameter space in which discoveries were previously expected.

Figure 2.5 summarizes the lower bounds on sparticle masses set by the analysis of Run I data at CMS and ATLAS. Pair-production of gluinos, first- and second-generation squarks, third-generation squarks, sleptons, and gauginos are considered in specific decay modes. Since the lower bounds on the masses of the pair-produced particles depend on the mass of the LSP (m_{LSP}), two cases $m_{LSP} = 200$ GeV and $m_{LSP} = 0$ GeV are presented in the figure. A range is given for each exclusion bound to account for the variety of mass lower limits obtained from multiple analysis techniques which are sensitive to different regions of the MSSM parameter space. Note that Fig. 2.5 is by no means an exhaustive summary of more than 100 SUSY searches conducted at the two experiments.

The gluino typically undergoes a three-body to quark–antiquark pair plus the LSP is when the squarks are heavy. In the analyses, t$\overline{t}$, b$\overline{b}$, and other decays are considered separately, but in all cases the gluino is generally excluded up to a mass of 1.2–1.4 TeV. If the initial production is instead a squark pair, the signature would be similar but with fewer jets because the squarks each decay to a quark and the LSP. Mass lower limits of ~850 GeV are set for the first- and second-generation squarks. For the third-generation squarks, the limit is somewhat weaker, at 600–700 GeV.

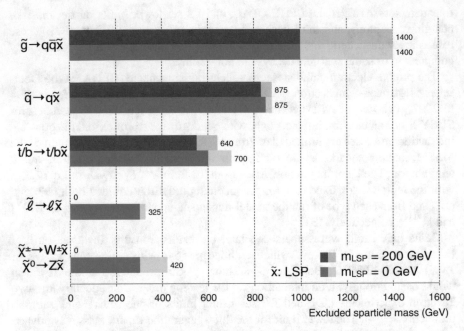

Fig. 2.5 A non-exhaustive summary of the results of SUSY searches from the LHC Run I, expressed in excluded sparticle mass ranges. For each pair-produced sparticle, the specific decay modes shown in the plot are considered. The first- and second-generation squarks are denoted as $\tilde{q}$. The meaning of the ranges displayed in faint colors are explained in the text. Both CMS [35–37] and ATLAS [38–41] results are included

With the large statistics already collected in Run I, the LHC experiments have also updated the limits on the slepton and gaugino masses. Searches for direct production of slepton-like particles decaying into a lepton and the LSP concluded that sleptons with mass lower than about 300 GeV likely do not exist for a light LSP. For the gauginos, there is a significant disagreement between ATLAS and CMS collaborations. While ATLAS excludes a mass for a degenerate neutralino and chargino that decay to Z and W bosons up to around 425 GeV, CMS sees a slight excess in the multi-lepton signal and therefore cannot exclude masses above 280 GeV.

2.12.2 GMSB Searches

The search results from the LHC experiments presented in Sect. 2.12.1 effectively probes GMSB-like signatures when m_{LSP} is assumed to be 0 GeV. However, even in GGM, not every sparticle is considered equally likely to be the NLSP, and some of the more likely scenarios are not addressed by the analyses mentioned above. In the remainder of this section, the status of the search for GMSB-motivated SUSY signal is summarized.

SUSY searches explicitly targeting GMSB scenarios already existed at LEP. At the time, however, only minimal GMSB was known as a viable GMSB model. Accordingly, the searches were performed in the signatures of single and diphoton plus missing energy or leptons plus missing energy. The single-photon search probed neutralino–gravitino production. Long-lived sleptons were also looked for in searches for a displaced lepton vertex and heavy charged particles. From these searches [42–45], the pair-production of neutralinos with a mass up to 105 GeV and a cross section above 20 fb; neutralino–gravitino production with the neutralino mass up to 210 GeV and a cross section above 100 fb; and sleptons with a mass up to 100 GeV regardless of their lifetime are excluded. The program of minimal GMSB searches in the photon plus missing energy channel was carried on at the Tevatron, with both CDF and D0 experiments assuming electroweak production of neutralinos and setting similar limits on the neutralino mass to the LEP results [46–48]. The Tevatron experiments probed delayed photons for the first time [49], as possible signals from long-lived bino-like neutralinos.

At the LHC, given that the strong-production ($\widetilde{g}\widetilde{g}$, $\widetilde{g}\widetilde{q}$, $\widetilde{q}\widetilde{q}$) is expected to dominate, neither ATLAS nor CMS consider the direct electroweak production of bino-like neutralinos anymore. The benchmark simplified model is instead similar to that for the gluino search mentioned above, but with the LSP replaced with a bino-like neutralino NLSP. In such models, the sensitivity to the signal is hardly affected by the neutralino mass, and therefore the search result is given in terms of the lower limit on the gluino mass. Both experiments exclude gluinos of mass below 1300 GeV [50, 51]. ATLAS additionally interpreted the search result in terms of the production of a wino-like neutralino–chargino pair that subsequently decays to bino-like neutralinos. Again the sensitivity is not dependent on the bino mass, and the winos of mass below 570 GeV are excluded [50]. The LHC searches are also not limited to prompt diphotons. Prompt sleptons are covered by the general slepton search already described. There are also dedicated analyses looking for displaced leptons and photons [52–54], exploiting the spectacular angular resolution of the calorimeters.

Following the null results in the searches for bino-like NLSP, the attention is shifting to other neutralino NLSP scenarios. Given the implication of the Higgs boson mass to the gluino mass discussed in Sect. 2.11, direct production of electroweakinos is especially actively searched for at the LHC. Particularly, after the discovery of the Higgs boson, higgsino searches in the $h\widetilde{G}$ channel exploiting the knowledge of the Higgs mass were conducted. As a result, higgsino NLSP in this decay mode is excluded up to a mass of approximately 300 GeV [55].

For the wino co-NLSP scenario, the gaugino mass exclusion limits for $m_{\mathrm{LSP}} = 0$ GeV shown in Fig. 2.5 seem to be directly interpretable as the lower bound on the co-NLSP mass. However, these limits are set assuming a decay BR of 100% for the decay $\widetilde{\chi}^0 \to Z\widetilde{G}$ and thus should be slightly scaled down to be regarded as limits on the pure wino mass. Searches with the signature of one photon, large hadronic activity, and E_T^{miss} can also probe strongly produced sparticles that decay into wino-like charginos and neutralinos. CMS has performed such a search [56] and excludes gluinos of mass below 775 GeV for a 375 GeV wino-like NLSP.

Searches of a wino-like NLSP in the photon–lepton channel have been performed by both ATLAS [57] and CMS [58] using pp collision data from the 7 TeV run. The analysis by ATLAS, using a larger data set than CMS, sets the lower limit on the wino and gluino masses of 221 and 619 GeV, respectively. The mass of the gluino is constrained under the assumption that it is pair-produced and decays to a quark, an antiquark, and a wino.

References

1. CMS Collaboration: Observation of a new boson at a mass of 125 GeV with the CMS experiment at the LHC. Phys. Lett. B **716**, 30–61 (2012). doi:10.1016/j.physletb.2012.08.021, arXiv: 1207.7235 [hep-ex], CMS-HIG-12-028, CERN-PH-EP-2012-220
2. ATLAS Collaboration: Observation of a new particle in the search for the standard model Higgs boson with the ATLAS detector at the LHC. Phys. Lett. B **716**, 1–29 (2012). doi:10.1016/j.physletb.2012.08.020, arXiv: 1207.7214 [hep-ex], CERN-PH-EP-2012-218
3. Langacker, P.: Structure of the standard model (1995). arXiv: hep-ph/0304186
4. Dine, M.: TASI lectures on the strong CP problem (2000). arXiv: hep-ph/0011376
5. Schmaltz, M., Tucker-Smith, D.: Little Higgs review. Ann. Rev. Nucl. Part. Sci. **55**, 229–270 (2005). doi:10.1146/annurev.nucl.55.090704.151502, arXiv: hep-ph/0502182 [hep-ph]
6. Baer, H., Tata, X.: Weak Scale Supersymmetry: From Superfields to Scattering Events. Cambridge University Press, Cambridge (2006)
7. LaBelle, P.: Supersymmetry DeMYSTiFied. McGraw-Hill Education, New York (2009). ISBN:9780071636421
8. Martin, S.P.: A Supersymmetry primer. Adv. Ser. Direct. High Energy Phys. **21**, 1–153 (2010). doi:10.1142/9789814307505_0001, arXiv: hep-ph/9709356 [hep-ph]
9. Shirman, Y.: TASI 2008 Lectures: Introduction to Supersymmetry and Supersymmetry Breaking (2009). arXiv: 0907.0039 [hep-ph]
10. Barbieri, R., Hall, L.J., Strumia, A.: Violations of lepton flavor and CP in supersymmetric unified theories. Nucl. Phys. B **445**, 219–251 (1995). doi:10.1016/0550-3213(95)00208-A, arXiv: hep-ph/9501334 [hep-ph], IFUP-TH-72-94, UCB-PTH-94-29, LBL-36381
11. Hisano, J., Moroi, T., Tobe, K., Yamaguchi, M.: Lepton flavor violation via right-handed neutrino Yukawa couplings in supersymmetric standard model. Phys. Rev. D **53**, 2442–2459 (1996). doi:10.1103/PhysRevD.53.2442, arXiv: hep-ph/9510309 [hep-ph], TIT-HEP-304, NSF-ITP-95-127, KEK-TH-450, LBL-37816, UT-727, TU-491
12. Meade, P., Seiberg, N., Shih, D.: General gauge mediation. Prog. Theor. Phys. Suppl. **177**, 143–158 (2009). doi:10.1143/PTPS.177.143, arXiv: 0801.3278 [hep-ph]
13. Draper, P., Meade, P., Reece, M., Shih, D.: Implications of a 125 GeV Higgs for the MSSM and low-scale SUSY breaking. Phys. Rev. D **85**, 095007 (2012). doi:10.1103/PhysRevD.85.095007, arXiv: 1112.3068 [hep-ph]
14. Choudhury, S.R., Gaur, N.: Dileptonic decay of B_s meson in SUSY models with large tan Beta. Phys. Lett. B **451**, 86–92 (1999). doi:10.1016/S0370-2693(99)00203-8, arXiv: hep-ph/9810307 [hep-ph]
15. Babu, K., Kolda, C.F.: Higgs mediated $B^0 \rightarrow \mu^+\mu^-$ in minimal supersymmetry. Phys. Rev. Lett. **84**, 228–231 (2000). doi:10.1103/PhysRevLett.84.228, arXiv: hep-ph/9909476 [hep-ph], OSU-HEP-99-10, LBNL-44284, UCB-PTH-99-43, LBL-44284
16. Turner, M.S., Wilczek, F.: Positron line radiation from halo WIMP annihilations as a dark matter signature. Phys. Rev. D **42**, 1001–1007 (1990). doi:10.1103/PhysRevD.42.1001, FERMILAB-PUB-89-044-A

17. Pierpaoli, E., Borgani, S., Masiero, A., Yamaguchi, M.: The formation of cosmic structures in a light gravitino dominated universe. Phys. Rev. D **57**, 2089–2100 (1998). doi:10.1103/PhysRevD.57.2089, arXiv: astro-ph/9709047 [astro-ph], TUM-HEP-288-97, SFB-375-206

18. Rychkov, V.S., Strumia, A.: Thermal production of gravitinos. Phys. Rev. D **75**, 075011 (2007). doi:10.1103/PhysRevD.75.075011, arXiv: hep-ph/0701104 [hep-ph], IFUP-TH-07-1

19. Moroi, T., Murayama, H., Yamaguchi, M.: Cosmological constraints on the light stable gravitino. Phys. Lett. B **303**, 289–294 (1993). doi:10.1016/0370-2693(93)91434-O, TU-424

20. Beenakker, W., Hopker, R., Spira, M.: PROSPINO: a program for the production of supersymmetric particles in next-to-leading order QCD (1996). arXiv: hep-ph/9611232

21. Ruderman, J.T., Shih, D.: General neutralino NLSPs at the early LHC. J. High Energy Phys. **1208**, 159 (2012). doi:10.1007/JHEP08(2012)159, arXiv: 1103.6083 [hep-ph]

22. Nath, P.: Supersymmetry after the Higgs (2015). arXiv: 1501.01679 [hep-ph]

23. MEG collaboration: new constraint on the existence of the $\mu^+ \to e^+\gamma$ decay. Phys. Rev. Lett. **110**, 201801 (2013). doi:10.1103/PhysRevLett.110.201801, arXiv: 1303.0754 [hep-ex]

24. CMS and LHCb Collaborations: Observation of the rare $B_s \to \mu^+\mu^-$ decay from the combined analysis of CMS and LHCb data (2014). arXiv: 1411.4413 [hep-ex]

25. Stöckinger, D.: The muon magnetic moment and supersymmetry. J. Phys. B **34**, R45–R92 (2007). doi:10.1088/0954-3899/34/2/R01, arXiv: hep-ph/0609168 [hep-ph], EDINBURGH-2006-22

26. Bernabei, R., et al.: Final model independent result of DAMA/LIBRA-phase1. Eur. Phys. J. C **73**(12), 2648 (2013). doi:10.1140/epjc/s10052-013-2648-7, arXiv: 1308.5109 [astro-ph.GA]

27. CoGeNT Collaboration: Results from a search for light mass dark matter with a P-type point contact germanium detector. Phys. Rev. Lett. **106**, 131301 (2011). doi:10.1103/PhysRevLett.106.131301, arXiv: 1002.4703 [astro-ph.CO]

28. SuperCDMS Collaboration: Search for low-mass weakly interacting massive particles with Supercdms. Phys. Rev. Lett. **112**(24), 241302 (2014). doi:10.1103/PhysRevLett.112.241302, arXiv: 1402.7137 [hep-ex]

29. Fermi LAT Collaboration: Measurement of separate cosmic-ray electron and positron spectra with the Fermi large area telescope. Phys. Rev. Lett. **108**, 011103 (2012). doi:10.1103/PhysRevLett.108.011103, arXiv: 1109.0521 [astro-ph.HE]

30. PAMELA Collaboration: Cosmic-ray positron energy spectrum measured by PAMELA. Phys. Rev. Lett. **111**(8), 081102 (2013). doi:10.1103/PhysRevLett.111.081102, arXiv: 1308.0133 [astro-ph.HE]

31. AMS Collaboration: High statistics measurement of the positron fraction in primary cosmic rays of 0.5–500 GeV with the alpha magnetic spectrometer on the international space station. Phys. Rev. Lett. **113**(12), 121101 (2014). doi:10.1103/PhysRevLett.113.121101

32. Lipniacka, A.: Understanding SUSY limits from LEP (2002). arXiv: hep-ph/0210356

33. Brandt, G.: Recent HERA results sensitive to SUSY (2008). arXiv: 0809.3509 [hep-ex]

34. CDF and D0 Collaborations: SUSY searches at the Tevatron. EPJ Web Conf. **28**, 09006 (2012). doi:10.1051/epjconf/20122809006, arXiv: 1202.0712 [hep-ex], FERMILAB-CONF-12-028-E, LAL-11-374

35. CMS Collaboration: Search for supersymmetry in hadronic final states using M_{T2} with the CMS detector at $\sqrt{s} = 8$ TeV. Tech. rep. CMS-PAS-SUS-13-019 (2014)

36. CMS Collaboration: Search for direct production of bottom squark pairs. Tech. rep. CMS-PAS-SUS-13-018 (2014)

37. CMS Collaboration: Searches for electroweak production of charginos, neutralinos, and sleptons decaying to leptons and W, Z, and Higgs bosons in pp collisions at 8 TeV. Eur. Phys. J. C **74**(9), 3036 (2014). doi:10.1140/epjc/s10052-014-3036-7, arXiv: 1405.7570 [hep-ex]

38. ATLAS Collaboration: Search for direct production of the top squark in the all-hadronic ttbar + etmiss final state in 21 fb^{-1} of pp collisions at $\sqrt{s} = 8$ TeV with the ATLAS detector. Tech. rep. ATLAS-CONF-2013-024, ATLAS-COM-CONF-2013-011 (2013)

39. ATLAS Collaboration: Search for direct top squark pair production in final states with one isolated lepton, jets, and missing transverse momentum in $\sqrt{s} = 8\,\text{TeV}$ pp collisions using $21\,\text{fb}^{-1}$ of ATLAS data. Tech. rep. ATLAS-CONF-2013-037, ATLAS-COM-CONF-2013-038 (2013)

40. ATLAS Collaboration: Search for direct production of charginos, neutralinos and sleptons in final states with two leptons and missing transverse momentum in pp collisions at $\sqrt{s} = 8\,\text{TeV}$ with the ATLAS detector. J. High Energy Phys. **1405**, 071 (2014). doi:10.1007/JHEP05(2014)071. arXiv: 1403.5294 [hep-ex], CERN-PH-EP-2014-037

41. ATLAS Collaboration: Search for squarks and gluinos with the ATLAS detector in final states with jets and missing transverse momentum using $\sqrt{s} = 8\,\text{TeV}$ proton–proton collision data. J. High Energy Phys. **1409**, 176 (2014). doi:10.1007/JHEP09(2014)176, arXiv: 1405.7875 [hep-ex], CERN-PH-EP-2014-093

42. ALEPH Collaboration: Search for gauge mediated SUSY breaking topologies in e^+e^- collisions at center-of-mass energies up to 209 GeV. Eur. Phys. J. C **25**, 339 (2002). doi:10.1007/s10052-002-1005-z, arXiv: hep-ex/0203024 [hep-ex], CERN-EP-2002-021

43. DELPHI Collaboration: Photon events with missing energy in e^+e^- collisions at $\sqrt{s} = 130\,\text{GeV}$ to 209 GeV. Eur. Phys. J. C **38**, 395 (2005). doi:10.1140/epjc/s2004-02051-8, arXiv: hep-ex/0406019 [hep-ex], CERN-EP-2003-093

44. L3 Collaboration: Single photon and multiphoton events with missing energy in e^+e^- collisions at LEP. Phys. Lett. B **587**, 16 (2004). doi:10.1016/j.physletb.2004.01.010, arXiv: hep-ex/0402002 [hep-ex], CERN-EP-2003-068

45. OPAL Collaboration: Searches for gauge-mediated supersymmetry breaking topologies in e^+e^- collisions at LEP2. Eur. Phys. J. C **46**, 307 (2006). doi:10.1140/epjc/s2006-02524-8, arXiv: hep-ex/0507048 [hep-ex], CERN-PH-EP-2005-025

46. CDF and D0 Collaborations: Searches for supersymmetry at the Tevatron. Springer Proc. Phys. **108**, 144–148 (2006). doi:10.1007/978-3-540-32841-4_28, FERMILAB-CONF-05-439-E

47. CDF Collaboration: Search for anomalous production of diphoton events with missing transverse energy at CDF and limits on gauge-mediated supersymmetry-breaking models. Phys. Rev. D **71**, 031104 (2005). doi:10.1103/PhysRevD.71.031104, arXiv: hep-ex/0410053 [hep-ex], FERMILAB-PUB-04-299-E

48. D0 Collaboration: Search for supersymmetry in di-photon final states at $\sqrt{s} = 1.96\,\text{TeV}$. Phys. Lett. B **659**, 856–863 (2008). doi:10.1016/j.physletb.2007.12.006, arXiv: 0710.3946 [hep-ex], FERMILAB-PUB-07-560-E

49. CDF Collaboration: Search for heavy, long-lived neutralinos that decay to photons at CDF II using photon timing. Phys. Rev. D **78**, 032015 (2008). doi:10.1103/PhysRevD.78.032015, arXiv: 0804.1043 [hep-ex], FERMILAB-PUB-08-078-E

50. ATLAS Collaboration: Search for diphoton events with large missing transverse momentum in 8 TeV pp collision data with the ATLAS detector. Tech. rep. ATLAS-CONF-2014-001, ATLAS-COM-CONF-2013-128 (2014)

51. CMS Collaboration: Search for supersymmetry in two-photons+jet events with razor variables in pp collisions at $\sqrt{s} = 8\,\text{TeV}$. Tech. rep. CMS-PAS-SUS-14-008 (2014)

52. ATLAS Collaboration: Search for nonpointing and delayed photons in the diphoton and missing transverse momentum final state in 8 TeV pp collisions at the LHC using the ATLAS detector. Phys. Rev. D **90**(11), 112005 (2014). doi:10.1103/PhysRevD.90.112005, arXiv: 1409.5542 [hep-ex]

53. ATLAS Collaboration: Searches for heavy long-lived charged particles with the ATLAS detector in proton-proton collisions at $\sqrt{s} = 8\,\text{TeV}$. J. High Energy Phys. **1501**, 068 (2015). doi:10.1007/JHEP01(2015)068, arXiv:1411.6795 [hep-ex], CERN-PH-EP-2014-252

54. CMS Collaboration: Search for long-lived particles decaying to photons and missing energy in proton-proton collisions at $\sqrt{s} = 7\,\text{TeV}$. Phys. Lett. B **722**, 273–294 (2013). doi:10.1016/j.physletb.2013.04.027, arXiv: 1212.1838 [hep-ex], CMS-EXO-11-035, CERN-PH-EP-2012-342

55. CMS Collaboration: Searches for electroweak neutralino and chargino production in channels with Higgs, Z, and W bosons in pp collisions at 8 TeV. Phys. Rev. D **90**(9), 092007 (2014). doi:10.1103/PhysRevD.90.092007, arXiv: 1409.3168 [hep-ex]
56. CMS Collaboration: Search for supersymmetry in events with one photon, jets and missing transverse energy at $\sqrt{s} = 8$ TeV. Tech. rep. CMS-PAS-SUS-14-004 (2014)
57. ATLAS Collaboration: Search for supersymmetry in events with at least one photon, one lepton, and large missing transverse momentum in proton–proton collision at a center-of-mass energy of 7 TeV with the ATLAS detector. ATLAS Conference Note 2012-144. CERN (2012)
58. CMS Collaboration: Search for supersymmetry in events with a lepton, a photon, and large missing transverse energy in pp collisions at $\sqrt{s} = 7$ TeV. J. High Energy Phys. **1106**, 093 (2011). doi:10.1007/JHEP06(2011)093, arXiv: 1105.3152 [hep-ex], CERN-PH-EP-2011-058, CMS-SUS-11-002

Chapter 3
The LHC and the CMS Experiment

3.1 The Large Hadron Collider

The CERN Large Hadron Collider (LHC) is the world's largest accelerator, storage ring, and proton–proton collider. Its two orbits are filled by oppositely circulating particle beams which collide at four interaction points (IP). Located at the interaction points are the particle detectors designed for different physics programs: A Toroidal LHC Apparatus (ATLAS) and Compact Muon Solenoid (CMS), general-purpose detectors for searches of physics beyond the standard model and precision measurements of QCD and electroweak interactions; A Large Ion Collider Experiment (ALICE), specializing in hadron physics; LHC beauty (LHCb), for precision heavy-flavor physics; and Large Hadron Collider forward (LHCf) and TOTEM (Total Cross Section, Elastic Scattering, and Diffraction Dissociation Measurement at the LHC), for hadron physics in the forward region and the proton cross section measurement.[1] Note that LHC is designed to also circulate lead ion ($^{208}\text{Pb}^{82}$) beams, and the ion–ion and ion–proton collisions are important parts of the LHC physics program. However, the remainder of this thesis will focus on the proton–proton operation.

Geographically, the LHC is placed in a near-circular tunnel which housed the Large Electron–Positron Collider (LEP) until its shutdown in 2000. The tunnel is 45–170 m beneath the ground passing the Franco-Swiss border line, with three quarters of its circumference under French territory. Being the latest addition to the CERN accelerator complex, the LHC draws its beams from a chain of pre-stage accelerators which were, and for some of them still are, at the frontier of high-energy physics programs. The schematic of the accelerator complex is shown in Fig. 3.1. The proton beams originate in the Linear accelerator 2 (Linac 2) and are brought

---

[1] There is a seventh experiment, MoEDAL, to search for heavy stable particles planned to start operation in the near future.

© Springer International Publishing AG 2017
Y. Iiyama, *Search for Supersymmetry in pp Collisions at $\sqrt{s} = 8$ TeV with a Photon, Lepton, and Missing Transverse Energy*, Springer Theses,
DOI 10.1007/978-3-319-58661-8_3

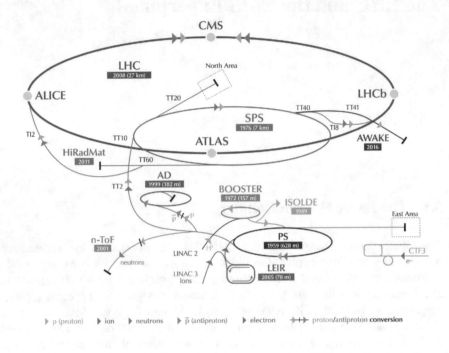

Fig. 3.1 Schematic view of the CERN accelerator complex. Retrieved from CERN Document Server [1]

to an energy of 50 MeV. They are then passed to the Proton Synchrotron Booster and to the Proton Synchrotron (PS), which provides acceleration to 1.4 and 25 GeV, respectively. After leaving the PS, the beams are further accelerated to 450 GeV in the Super Proton Synchrotron (SPS) whose circumference passes through the two CERN sites, Meyrin and Prévessin, and are finally fed to the LHC.

The maximum beam energy of the LHC is 7 TeV, designed to deliver pp collisions at $\sqrt{s} = 14$ TeV at the interaction points. Its target instantaneous luminosity is 10 nb^{-1} s^{-1}, which is a beam intensity that corresponds to the production of roughly 10,000 W bosons, 100 $t\bar{t}$ pairs, and five Higgs bosons [2] every 10 s at the maximum energy. Taking into account the machine operational efficiency, fill length, required intervals between the fills, and a few technical stops per year, the machine at its design specification can deliver about 100 fb^{-1} of integrated luminosity per year per interaction point. The desire for high luminosity was one of the reasons that a pp machine was chosen over a p$\bar{p}$ machine like the Fermilab Tevatron; protons are extremely cheaply obtainable by stripping the electrons off hydrogen molecules, while antiprotons need to be produced in particle inelastic collisions, stored, and accumulated. A small hydrogen tank at the source end of Linac 2 can provide protons for years of LHC running.

The upper bound on the beam energy is dictated by the tunnel geometry and the attainable dipole magnetic field to bend the beam paths. The formula that relates the magnetic field B to the radius of curvature R of the trajectory of a charged particle with momentum p and charge q is

$$B = \frac{p}{qcR},$$ (3.1)

where c is the speed of light. The tunnel geometry defines the minimum R in the beam orbit. In fact, since electrons and positrons of the LEP required multiple accelerations per orbit to compensate for the energy loss due to the synchrotron radiation, the tunnel has eight straight sections where the accelerating RF cavities were placed. Consequently, the bend of the beam path at the curved sections of the tunnel is tighter than what it would be in a perfectly circular ring. Thus the maximum attainable beam energy for the LHC could be higher for a given magnetic field if the tunnel was rebuilt, but the cost calculation demanded the reuse of the LEP tunnel. The maximum magnetic field of the LHC dipoles is $B = 8.33$ T.

The boundary condition of reusing the LEP tunnel has in fact affected many other design parameters of the LHC. The most notable is the adoption of the "twin-bore" design of the magnets, i.e., embedding two beam pipes into a single magnet system. The decision was driven by the need for the bending magnets to be superconducting to achieve the necessary field strength. The compact interior of the tunnel would not allow the massive insulation and helium distribution systems required by the superconducting magnets to be independently built for two beam rings. The only viable solution was therefore the twin-bore design, where the two beam pipes are placed next to each other in a common cold mass, immersed in the return field of each other's dipole field. This design is actually cost-effective compared to doubling the number of cold masses, but has a disadvantage in the flexibility of beam control.

The dipole design also defines the beam pipe aperture and with it the maximum size of the beam itself. Using this and the peak beta function given by the tunnel geometry, a bound for the allowed beam emittance can be obtained. It is known that there is a theoretical maximum to the ratio of the number of protons per bunch N_b to the emittance, beyond which the beam–beam interaction becomes too strong. Employing an empirical value for this maximum, the design N_b of the LHC was determined to be 1.15×10^{11}. From this follows that the target instantaneous luminosity is achievable with a bunch spacing of 25 ns, or equivalently a bunch crossing frequency of 40 MHz.

As already mentioned, the dipole magnets, of which there are 1236 in the LHC ring, are the most critical components of the LHC. They operate at 1.9 K, cooled by superliquid helium. To limit the amount of mass that is cooled down to this low temperature, the solenoids are surrounded by multiple insulating layers of different temperatures. The dipole container, triumphantly displayed on the lawn in front of Restaurant 1 at CERN, is also vacuumized in order to eliminate convection inside. The solenoid is made of stacked niobium-titanium filaments of thickness 6–7 μm.

The LHC has only one accelerating sector. In this sector, there are eight superconducting RF cavities per beam operated at 400 MHz. During acceleration, the cavities add 485 keV to each proton per turn. The energy lost from the protons due to the synchrotron radiation is 0.1–7000 eV per turn per proton, and is thus negligible.

After an accident in 2008 which damaged the dipole magnets in one ring sector, the LHC started its Run I physics program in 2010 with the reduced beam energy of 3.5 TeV. With confidence gained through the 2-year operation, in 2012, which was the last year of Run I, the beam energy was raised to 4 TeV. The bunch crossing frequency during Run I was reduced to 20 MHz, but N_b at the end of 2012 was already slightly higher than the design value. Figure 3.2 is a graph showing the evolution of integrated luminosity at the CMS experiment in 2012. In total, CMS recorded 21.79 fb^{-1} of pp data in 2012.

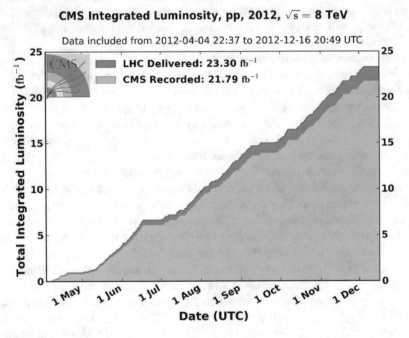

Fig. 3.2 Evolution of pp integrated luminosity at CMS during the 2012 data taking. Retrieved from [3]

3.2 The CMS Detector

CMS is a hermetic, general-purpose detector built with the main goal of discovering unknown particles and high-energy phenomena as well as studying the SM at the LHC. Its central feature is a superconducting solenoid of 6 m internal diameter, which provides a highly uniform magnetic field of 3.8 T within its bore. This strong magnetic field allows a smaller detector volume than, e.g., ATLAS. The full dimensions of the CMS detector with the iron return yoke of the magnet and the outer detectors are 21.6 m in the axial direction and 14.6 m in diameter. The overall layout is illustrated in Fig. 3.3.

The superconducting magnet is a 4-layer niobium-titanium coil embedded in aluminum and aluminum alloy. The conducting mass is self-supporting to resist the extreme magnetic pressure at full current. The entire conductor is cooled down to 4.5 K by liquid helium and is suspended in a vacuum cryostat for insulation. Figure 3.4 is an image of the computed field strength and field lines within CMS, showing a high field uniformity in the bore of the solenoid.

The coordinate system used in CMS is suited to the overall cylindrical structure of the detector. The axis of the cylinder, along which the beam pipe runs, is defined as the z-axis, with the positive direction pointing along beam 2 (rotating counter-clockwise as seen from the sky) of the LHC. The system is right-handed, and the x- and y-axes point to the center of the LHC and upward, respectively. The coordinate

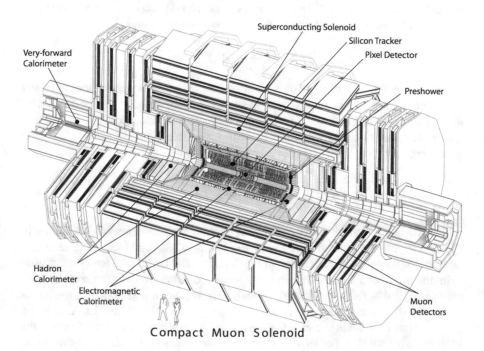

Fig. 3.3 Perspective view of the CMS detector. Reprinted from [4]

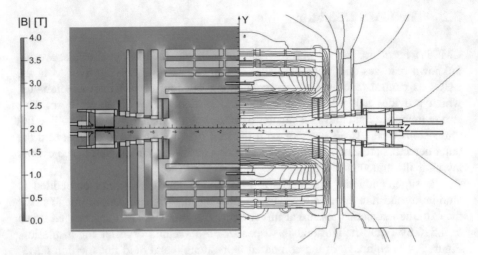

Fig. 3.4 Computed field strength and field lines on a longitudinal section of the CMS detector. Each field line represents a magnetic flux increment of 6 Wb. The region at the center corresponds to the bore of the solenoid. Reprinted from [5]

origin is at the geometrical center of the cylinder. The radial distance $r = \sqrt{x^2 + y^2}$ and the azimuthal angle $\phi = \arctan(y/x)$ are often used instead of x and y, reflecting the cylindrical symmetry. The momenta of the physics objects in the transverse plane are also expressed in terms of the transverse momentum

$$\vec{p_T} = (p_x, p_y) \tag{3.2}$$

and ϕ. The magnitude of $\vec{p_T}$ is often simply denoted p_T. When calorimetric aspect of the physics object is emphasized, the transverse energy $E_T = E \sin \theta$ is also used. The polar angle θ is defined by $\tan \theta = r/z$. It also should be noted that kinematic quantities involving the z-direction are almost always expressed by the pseudorapidity

$$\eta = -\log \tan \left(\frac{\theta}{2} \right). \tag{3.3}$$

The CMS detector consists of a barrel section and two endcaps, as shown in Fig. 3.3. Figure 3.5 is an illustration of the layout of the CMS detector, seen in a quadrant of the longitudinal section of the detector. The barrel consists of concentric cylindrical detector layers. The inner detector contained within the magnet bore comprises, going radially outward: three layers of silicon pixel tracker; ten layers of silicon strip tracker; a lead-crystal electromagnetic calorimeter (ECAL); and a plastic scintillator hadronic calorimeter (HCAL) with brass absorber. Outside the solenoid are: outer layer of HCAL (HO); iron magnetic return yokes; and drift tubes (DT) and resistive plate chambers (RPC) interlayered between the return yokes.

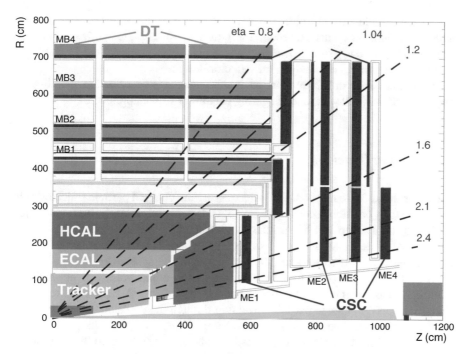

Fig. 3.5 Longitudinal layout of one quadrant of the CMS detector. The volume labeled as "Tracker" contains both pixel and strip trackers. HO and HF are not labeled in the image. Reprinted from [6]

The endcaps are mirror images of each other and are also layered. The ordering of the detector types is identical to that of the barrel, except for a preshower detector, which is placed in between the silicon strip tracker and the ECAL, while cathode strip chambers (CSC) replace the DT. There are also forward calorimeters (HF) further away from the endcaps that detect low-angle scatterings. The different detector layers are often referred to as sub-detectors.

As a multi-purpose detector, the design goal of CMS is to achieve the best particle identification and 4-momentum reconstruction possible with the available technology. The sub-detectors must all work complementarily and coherently for this goal. In addition to technological difficulties, there is a challenge posed by the high luminosity of the LHC, resulting in multiple pp interactions within a single bunch crossing. Since an overwhelming majority of pp collisions is uninteresting low-energy scattering, a typical event of interest has one hard-scattering interaction and many other interactions from which low-energy tracks and radiations emerge. The number of such additional soft-scattering interactions, called pileups (PU), ranged from 5 to 40 for Run I.

It is also important to note that an interesting hard-scattering interaction does not occur at every bunch crossing. This is in a sense fortunate, since it is also impossible to read out and record the full detector data at 40 MHz. In fact, CMS has a highly sophisticated trigger system (see Sect. 3.3) that selects a few hundred interesting collision events per second for recording. Since parts of this system must be hardware-based to process detector data as quickly as possible, the CMS trigger was one of the central elements in the detector design.

In the remainder of this section, each sub-detector is described in detail on its operating principle, layout, and triggering capability where applicable.

3.2.1   Inner Trackers

The inner trackers are responsible for reconstructing charged particle trajectories to determine their position according to Eq. (3.1). Equally importantly, high-resolution tracks can be used to identify primary and secondary vertices, which are crucial for PU identification and heavy-flavor reconstruction. The inner tracker volume in CMS extends to 1.2 m in r and 2.9 m in z. All of the sensors of the tracker are made of silicon semiconductors. In a silicon sensor, a bias voltage is applied to a volume around the junction of two types of semiconductors to deplete the region of charge carriers. A charged particle traversing through the depleted region produces electron–hole pairs, which are subsequently collected as signal charge. Thus the operational principle of silicon detectors is fundamentally similar to that of gaseous wire chambers, which exploits gas ionization, but the former can provide orders of magnitude higher spacial resolution.

3.2.1.1   Pixel Tracker

The pixel tracker is the detector closest to the interaction region of CMS. The sensors of the pixel tracker are silicon 8 mm × 8 mm squares segmented in 100 μm × 150 μm pixels. The silicon squares are aligned side-by-side to form planer modules, which are then arranged into three cylinders for the barrel and four disks for the endcap. There are approximately 48 million independent readout channels in the pixel tracker system. The barrel cylinders have radii of 4.4, 7.3, and 10.2 cm, and the endcap disks are located at $z = \pm 34.5$ and ± 46.5 cm. Figure 3.6 shows the detailed layout of the inner trackers.

The geometry of the pixel tracker is such that the tracks cross the sensor planes with inclinations on average 20° from normal. This causes the charge deposit from a single track to be shared among multiple pixels, an effect further enhanced by the Lorentz drift. CMS takes advantage of this charge-sharing with the analog readout of the collected charge. An interpolation of the signal amplitudes gives a spacial resolution of 15–20 μm, which is much higher than the raw pixel spacing. The price to pay is the increased data volume, which does not allow a serial readout as is done in charge-coupled devices. Therefore, the data from the individual pixels are read out directly by the readout chip bump-bonded behind the sensors.

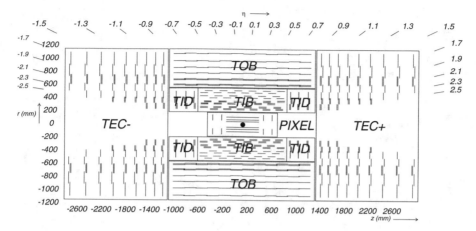

Fig. 3.6 The layout of the inner tracker detectors. The volumes labeled TIB, TID, TOB, and TEC are all strip trackers. *TIB* Tracker Inner Barrel, *TID* Tracker Inner Disk, *TOB* Tracker Outer Barrel, *TEC* Tracker Endcap. Reprinted from [4]

3.2.1.2 Strip Tracker

The strip tracker covers the remainder of the inner tracking volume. It also consists of planer silicon sensors arranged into cylindrical and disk-like structures. There are in total 10 layers of cylinders for the barrel and 12 layers of disks for each of the endcaps. Unlike the pixel tracker, typical dimension of a strip sensor "cell" is $10\,\mathrm{cm} \times 80\,\mu\mathrm{m}$. Thus its sensor panels feature arrays of fine strips aligned in one direction with narrow spacing, or pitch. Strips in the barrel run along z-direction with the pitch ranging from 80 to 183 μm. The smallest pitch is present in the innermost layer. In the endcaps, the strips are aligned radially with the pitch ranging from 100 to 184 μm.

From its geometry, it is obvious that the strip tracker provides a fine resolution in $r-\phi$ for barrel and $z-\phi$ for the endcap, but very little information on the orthogonal directions unless they are combined with non-parallel strips. Therefore, more than a third of the strip tracker modules are actually double-layered, with the second layer tilted by 100 mrad with respect to the first, allowing a stereo measurement. The full spacial resolution of the strip tracker is thus 25–50 μm in the direction perpendicular to the strips and 230–530 μm along the strips on the stereo modules.

The alignment of the large structures, i.e., groups of layers, of the strip tracker is constantly monitored by a laser alignment system. An infrared laser light is distributed and shines directly onto specific sensors to induce signal pulses. The system can run at 100 Hz and completes a full cycle in a few seconds. The alignment data is taken not only during commissioning and inter-fill periods, but also during physics runs using the beam abort gap. An alignment precision of 100 μm is attainable with this system.

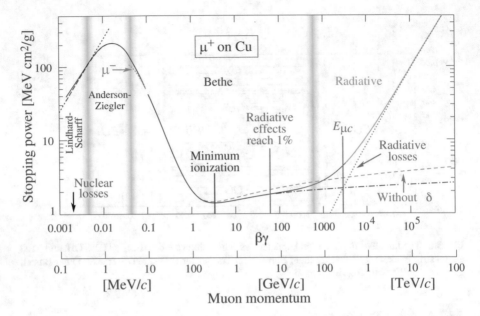

Fig. 3.7 Stopping power for high-energy μ^+ in copper. The qualitative behavior is similar for other energetic charged particles, and is mostly dependent on the relativistic $\beta\gamma$ factor. Reprinted from [7] (32. Passage of particles through matter)

3.2.2   Electromagnetic Calorimeter and Preshower Detector

The electromagnetic calorimeter (ECAL) measures the energy of electrons and photons by means of full energy absorption. Electrons with energy above a few GeV deposit energy in a dense material mostly by radiation (bremsstrahlung). This behavior can be understood from Fig. 3.7, which plots out the stopping power curve of μ^+ in copper. The qualitative feature of the interaction of high-energy charged particles in matter is known to be dependent only on the relativistic $\beta\gamma$ factor. From the figure, radiative energy loss dominates above $\beta\gamma \sim 1000$. Since $\beta\gamma$ is equal to the momentum-to-mass ratio of the particle, this threshold translates to a momentum of a few GeV for electrons, while it is much higher for other particles. An energetic photon also interacts in a dense medium by an e^+e^- pair production. Thus, in effect, a high-energy electron or photon that impinges on a dense material loses energy rapidly, generating one another in turns, causing an electromagnetic shower. The main design consideration of the ECAL in the LHC detectors is therefore to have a mass volume dense and deep enough to capture electrons and photons with energies up to a few TeV.

The absorbing material of the CMS ECAL is crystalline lead tungstate ($PbWO_4$), which is also a scintillator. The total energy of an electromagnetic shower can be inferred by collecting the scintillation light it generates. Therefore, a homogenous volume of $PbWO_4$ acts both as an absorber and an active detector material.

The intensity of the scintillation light, which scales approximately linearly with the shower energy deposit, is measured by photosensors directly glued to the crystals, and is converted into an electric signal.

The ECAL is a single-layer detector, consisting of arrays of rod-shaped crystals aligned side by side. The ECAL barrel (EB) is made of a row along the z-direction of 170 crystals repeated in a 360-fold symmetry along ϕ. The crystal row along z-direction covers $|\eta| < 1.479$. The crystals have a truncated pyramidal shape. The front face, i.e., the face that is closest to the tracker, of an EB crystal has a 22 mm $\times$ 22 mm cross section, while at the rear face it is 26 mm $\times$ 26 mm. The front face cross section roughly corresponds to 0.0174×0.0174 in $\Delta\eta \times \Delta\phi$. The length of an EB crystal is 230 mm. The ECAL endcaps (EE) each has 7324 crystals organized in an x–y grid that covers a pseudorapidity range $1.479 < |\eta| < 3.0$. The EE crystals have more cuboid-like shapes, with the front and rear face dimensions of 28.6 mm $\times$ 28.6 mm and 30 mm $\times$ 30 mm, respectively. The crystal length in the EE is 220 mm.

The crystal lengths in the EB and EE, respectively, correspond to 25.8 and 24.7 radiation lengths. One radiation length is the density-weighted depth of material in which a high-energy electron loses in average all but e^{-1} of its initial energy due to radiation. Equivalently, it is $7/9$ of the mean free path for pair production by a high-energy photon. Thanks to this depth, most of the electrons and photons have no energy leakage beyond ECAL. In the lateral direction, the density of PbWO$_4$ implies a small Molière radius of 2.2 cm. The Molière radius is the average radius of a cylinder containing 90% of the shower energy. When an electron or a photon arrives at the ECAL front face without having already undergone bremsstrahlung and/or pair creation due to the tracker material, more than 90% of its energy is typically contained within a 3×3 crystal matrix around the point of incidence.

Figure 3.8 shows the layout of the full ECAL. The crystals are separated by thin walls to avoid light leakage (cross-talk), and organized in a hierarchical structure. Within a single EB submodule, which is the smallest structural unit consisting of eight crystals, the walls are made of aluminum and resin and measures 0.35 mm in thickness. The separations between the submodules, modules, and supermodules, which are the larger structural units, are thicker. The crystals in the EE are separated by similar spacings. The crystals do not point directly to the interaction region (have quasi-projective geometry) to reduce the likelihood of the primary photon or electron emerging from the hard scattering passing along the crystal walls, depositing most of its energy in a passive material.

The advantages of PbWO$_4$ as a scintillator compared to other inorganic scintillating crystals are a fast decay time of scintillation and radiation-hardness. The decay time constant of approximately 25 ns in principle allows discrimination between two signals arriving in contiguous bunch crossings. The radiation-hardness is a critical property for the ECAL to survive the high luminosity of the LHC for its years of operation. Inorganic scintillators are known to lose their transparency under severe radiation, due mainly to defects in the crystal lattice caused by nuclear interactions. Of several candidate materials considered for the ECAL, PbWO$_4$ was the most promising in terms of stability of the light transmission. The light output itself, or the amount of scintillation photon per energy deposit, of PbWO$_4$ is less than 1%

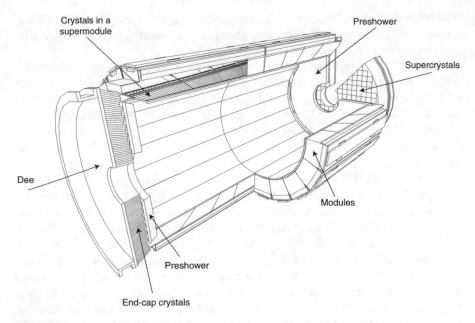

Fig. 3.8 The layout of the ECAL and preshower detector. Reprinted from [4]

compared to NaI, a common inorganic scintillator. However, this is not a critical disadvantage, since the typical energy of particles measured by the ECAL would still produce enough scintillation photons for the Poissonian error to be negligible.

The photosensors in the EB are avalanche photodiodes (APD). In this solid-state sensor, the photo-sensitive surface emits an electron when hit by a photon. The emitted electron signal is amplified through avalanche of ionizations in the region of the internal diode where a high reverse-bias voltage is applied (depletion layer). The APD used in the CMS EB is capable of multiplying the number of electrons by more than 1000-fold at the operating voltage of 340–430 V. The choice of APD as the photosensor is driven by the strong magnetic field and the EB geometry in which the photo-sensitive surface is nearly parallel to the field. A conventional photomultiplier is not operable in such an environment, since the electron flow within the multiplier tube would be disrupted by the magnetic field. In contrast, the depletion layer of the APD has an effective thickness of $6 \pm 0.5\,\mu$m, making the APD almost insensitive to the magnetic field. Two APDs are glued to the rear end of each crystal, collecting approximately 4.5 photoelectrons per MeV of energy deposit per sensor.

In the EE, vacuum phototriodes (VPT) are used as the photosensor. VPT is a photomultiplier tube with a single gain stage. In the EE geometry, the magnetic field is normal to the photocathodes of the VPTs, and thus has only a slight effect on the VPT gain. There is one VPT per crystal, also glued to the rear end. The output electric signal per shower energy in the EE is similar to that in the EB.

The small signals from the photodetectors are shaped and amplified in the Multi-Gain Preamplifier (MGPA) developed specifically for the CMS ECAL. The MGPA has gain modes 1, 6, and 12, where gains 1 and 6 are chosen for the output once the signal has saturated gains 6 and 12, respectively. This dynamic mechanism gives a signal dynamic range from a few MeV to 1.5 (1.6–3.1) TeV in the EB (EE) with a 12-bit analog-to-digital converter (ADC). The output voltage pulse has a length of approximately 300 ns, or 12 LHC clocks (bunch crossing times), where the pulse maximum is at the third clock and a slow decay follows. The ADC samples the signal at each clock.

An important function of ECAL besides providing a precise energy measurement of electrons and photons is to generate trigger primitives (TPs). A TP is a rough estimate of the energy deposit and its geometrical location measured at every bunch crossing. TPs generated in the ECAL are processed by the Regional Calorimeter Trigger (RCT), described in Sect. 3.3, together with the TPs from the HCAL, and used in the final Level-1 trigger decision. For the TP generation, multiple crystals are grouped into what are called trigger towers. At each clock, the ADC samples of the crystals in each trigger tower are summed up and converted to an energy estimation through a lookup table.

As already mentioned, the PbWO$_4$ crystals lose transparency over their lifetime. Since the number of photons corrected per GeV of shower energy will decrease, the apparent energy of electrons and photons will become lower in time, if no correction is applied. To monitor this effect, the crystals are continuously lit with laser light. The same laser beam is split and shines onto off-detector PN silicon photodiodes for reference measurements. The time evolution of the ratio of the laser signal strengths obtained from the ECAL and from the photodiodes is then used as input to the calculation of the transparency correction applied at signal reconstruction time. The laser irradiation takes place even during physics runs using the beam abort gaps, because there is a fast component to the radiation damage which creates an observable difference in transparency over a single LHC fill. Figure 3.9 shows the evolution of the crystal transparency during LHC Run I.

The basic calibration of the ECAL must address not just the stability of the detector response over time, but also its uniformity in space. After the transparency correction is applied to individual crystals, a potential crystal-to-crystal variation of the signal amplitude for a given energy deposit must be removed to simplify the particle energy reconstruction. This homogenization of relative detector response is called intercalibration. The CMS ECAL performed an in-situ intercalibration using collision data during Run I. A dedicated data stream was prepared to collect a large sample of $\pi^0 \to \gamma\gamma$ decays. Since the π^0 mass of 135 MeV is typically much smaller than its p_T in the LHC collisions, the two photons arrive at the ECAL nearly collinearly. Nevertheless, the high granularity of the CMS ECAL enables the distinction of the two photons, and thus allows to reconstruct the π^0 mass. The crystals were intercalibrated by adjusting the individual crystal responses iteratively so that the measured π^0 mass centers at its nominal value everywhere in the CMS detector.

The π^0 that is useful for intercalibration can be problematic for physics analyses based on identifying prompt photons. The primary purpose of the preshower detector in the endcaps is to force the initiation of an electromagnetic shower in

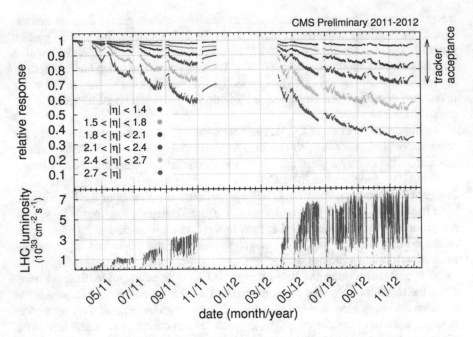

Fig. 3.9 The evolution of the mean crystal transparency normalized to the value at the beginning of Run I, displayed separately for different pseudorapidity ranges. Reprinted from [8]

a region with high spacial resolution before the ECAL, so that two nearly collinear photons can be disambiguated. It can also provide accurate measurements of the incident position of electrons. In CMS, the preshower detector consists of alternating layers of lead absorbers and silicon strip sensors, and is placed in front of the EE. The first layer of lead has a thickness corresponding to two radiation lengths. It is followed by a sensor plane with vertical strips of 6 cm individual length and 1.9 mm pitch. The second layer of lead is one radiation length thick, and is followed by another sensor plane with identical sensors oriented in the horizontal direction.

3.2.3   Hadronic Calorimeter and Forward Detectors

The Hadronic Calorimeter (HCAL) absorbs and measures the energy of particles that are not stopped by the ECAL. The particles that reach the HCAL are mainly ones heavier than the electron. Except for muons, all such particles are hadrons and thus undergo nuclear interaction with matter. It should be noted that nuclear interactions can already take place in the ECAL, which has a nuclear interaction length of 1.1. The nuclear interaction length is the mean free path for inelastic nuclear scattering. In the case of charged hadrons, further energy can be deposited in the ECAL through bremsstrahlung. Therefore, the HCAL inherently has a lower energy/momentum resolution compared to the inner trackers and the ECAL. The relative energy resolution of the CMS HCAL combined with the ECAL is roughly $\Delta E / E \approx 100\% / \sqrt{E\,[\text{GeV}]} \oplus 5\%$ [9].

Unlike ECAL, the HCAL is a heterogeneous calorimeter, i.e., the absorber and active detector materials are separated. Brass is used to induce hadronic shower, while the energy deposit of the showers is sampled using plastic scintillators. In the plastic volume, charged particles in the shower generate scintillation light, which is drawn out of the detector using wavelength-shifting optical fibers embedded into the scintillator. The fibers lead to photodetectors and electronics that are located outside the detector.

The HCAL Barrel (HB) fills the cylindrical volume from $r = 1.77$ m to 2.95 m and covers a pseudorapidity range $0 < |\eta| < 1.3$. Radially, it consists of 17 thin scintillator layers interleaved with thick absorber layers. The total absorber depth corresponds to 5.82 interaction lengths at $|\eta| = 0$ and 10.6 at $|\eta| = 1.3$. Each scintillator layer is segmented into small tiles. One tile covers an $\Delta \eta \times \Delta \phi$ area of 0.087×0.087, which is equivalent to a 5×5 EB crystal matrix. The tile boundaries are aligned radially, such that the HB active material can be seen to form towers in a quasi-projective geometry similar to the ECAL crystals. Indeed all signals from a single tower are read out in a single channel, except for the highest $|\eta|$ towers, which have two longitudinally segmented readout channels. In order to capture late or long showers, there are plastic scintillator tiles placed outside of the main solenoid, using the solenoid cold mass as an absorber. These tiles, called outer HCAL or HO, have coarser geometry than the HB tiles and act as "tail-catchers."

The HCAL Endcap (HE) covers the pseudorapidity range $1.3 < |\eta| < 3$ with a similar brass-scintillator tower layout. There are 18 scintillator and 17 absorber layers, with an interaction length of 10 when combined with the preshower detector and the EE. The HE tiles follow the same $\Delta \eta \times \Delta \phi$ granularity as the HE up to $|\eta| < 1.6$. The granularity is reduced to $\Delta \eta \times \Delta \phi = 0.17 \times 0.17$ for $|\eta| > 1.6$. Each tower is read out in two longitudinal segmentations, except for the highest pseudorapidity regions, where three segmentations are used.

The scintillation light collected by fibers is converted to electronic signals via hybrid photodiodes (HPDs). A HPD consists of a photocathode and multiple diode pixels. The photocathode is held at a large negative voltage of 8 kV. A photoelectron that is emitted from the cathode plane following a photon impingement is accelerated and hits the diode pixels. The diode pixels are placed under reverse-bias voltage, and just like the APDs used in the EB, the incoming electron causes an ionization avalanche in the depletion layer. The HPD is used for the HCAL because the photodetectors could not be placed outside of the magnetic field even when using fibers to carry the scintillation signal. The electronic pulse from the HPDs is digitized by charge-integrating ADCs into 7-bit data encoded with a non-linear scale to cover the required dynamic range.

The calibration of the scintillation detector system is carried out using Cs^{137} or Co^{60} radioactive sources. When CMS is not taking data, a small robot carrying a source can be deployed to travel through thin tubes that are embedded in the scintillator tiles. The source irradiates the tiles with γ rays of known energy, thus providing an absolute calibration.

Beyond $|\eta| > 3$, the forward HCAL (HF) supplements the detector coverage to $|\eta| < 5.2$. The absorber of the HF is made of steel. The active material is quartz fibers inserted into the absorber mass. The quartz fibers, chosen for their radiation-hardness, detect charged particles via Čerenkov radiation. The HF is calibrated together with the HB and HE using the source-carrying robot. There are also two ultra-high pseudorapidity detectors CASTOR and ZDC, which are used mainly in heavy-ion physics and measurements of diffractive proton scattering.

3.2.4   Muon Trackers

Muons are 200 times heavier than electrons and also do not interact strongly. Thus, most of the muons penetrate the calorimeters and the solenoid without significant energy loss. The muon trackers are placed in the outermost layers of CMS to detect muons and measure their momentum. They must also have trigger capability, since high-p_T muons are often a signature of interesting physics.

The CMS muon system is designed to cover the full pseudorapidity range up to $|\eta| < 2.4$. The detectors are located in what are called muon stations, which are spaces between and adjacent to the iron return yokes for the solenoid. Being outer detectors at $r > 3.5$ m, they must cover a large area, which makes silicon-based trackers an unrealistic and unnecessary choice. The large volume available outside the solenoid allows the usage of gaseous chambers to achieve a resolution comparable to that of silicon trackers. Indeed drift tubes and cathode strip chambers were chosen for the barrel and endcap muon tracking, respectively. In addition, resistive plate chambers are employed in both the barrel and endcap for enhanced triggering capability.

3.2.4.1   Drift Tubes

A drift tube is a type of wire drift chamber, which consists of a thin conducting wire at high positive voltage suspended in a gaseous volume. Ionization of the gas caused by a passing charged particle is amplified to an observable electric signal around the wire due to the electric field strength that is inversely proportional to the distance from the wire; electrons are accelerated more and more as they get closer to the wire. A drift chamber typically measures the time between the particle passage and the electron avalanche signal. Since the drift velocity is mainly dependent on the field configuration, gas type, and gas pressure, the distance to the point of initial ionization can be inferred from this time measurement. By utilizing multiple wires, the incident position of the particle along the direction perpendicular to the wires can be reconstructed with a good accuracy. The DT signal is measured by time-to-digital converter (TDC) with a 0.8 ns time resolution, using the LHC clock as time reference.

A DT is different from a simple multi-wire drift chamber in that each wire is housed in its own cell, or tube, and surrounded by lower-voltage electrodes. This design confines the damage inflicted by broken wires if there are any, and also prevents avalanche debris around one wire to affect the neighboring wires. Additionally, electrodes can be placed to achieve a field configuration with good linearity between the drift time and distance.

A single cell of the CMS drift tubes has a rectangular cross section with round corners. Cells are made in layers by separating a gap between two aluminum plates with parallel I-beams of 1 mm thickness. The gap height is 1.3 cm and the interval between I-beams is 4.2 cm. Cells are placed either parallel or perpendicular to the z-direction. For those that are parallel along z, the cell length is 2.4 m. A gold-plated stainless steel anode wire of 50 μm thickness runs lengthwise at the center of each cell.

The anode wire is held at 3.6 kV while the aluminum plates and I-beams are set to ground. Field electrodes made of aluminum tapes are glued on both plates and are set at 1.8 kV. The cathode aluminum tapes at −1.2 kV are glued on the I-beams. This rather complicated placement of electrodes realizes a nearly uniform electric field along the cell cross section, which is illustrated in Fig. 3.10. The DT uses a 85/15 mixture of Ar and CO_2 as the chamber gas.

The structural unit of the drift tubes is the superlayer, which is a staggered set of four cell layers. A DT chamber consists of two or three such superlayers. The chambers themselves follow layered structures. A chamber with three superlayers has one superlayer with the cells running parallel to z closest to the interaction point, followed by one low-density spacer, one superlayer with cells in perpendicular orientation, and another superlayer with cells along the z-direction. The chambers are arranged into dodecagonal rings with a 2.4 m thickness in the z-direction. There are four ring layers interleaved with iron yokes: One between the solenoid and the first iron layer, two inserted between the yokes, and one farthest from the interaction point. The DT ring layers correspond to the muon stations. The iron yokes and muon stations together form a CMS wheel. Five wheels stacked along the beam direction form the outer barrel. The full DT detector provides a pseudorapidity coverage of $0 < |\eta| < 1.2$.

The position resolution from each drift tube is approximately 250 μm. When information from other tubes in a chamber is combined, the resolution reaches 100 μm.

3.2.4.2 Cathode Strip Chambers

Cathode strip chambers are another type of multi-wire chambers. The CSC detector is made of panels parallel to the transverse plane. In each chamber, anode wires are spanned azimuthally, and the cathode, made of metallic strips, runs radially. Unlike the DT detector, the CSC does not measure the drift time. Instead, the particle position, in particular its ϕ coordinate, is measured with precision by interpolating the charge signal obtained in the cathode strips.

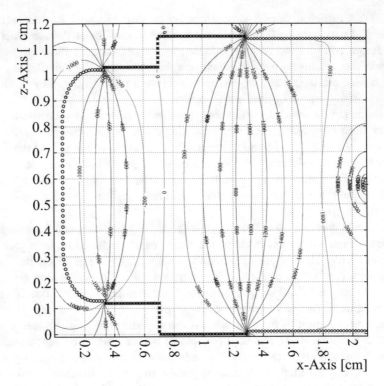

Fig. 3.10 Equipotential lines in half of a drift cell of the CMS drift tubes. The anode wire is located at the center of the right edge of the figure. *Small circles*, *asterisks*, and *crosses* represent the cathode tape, the grounded aluminum structure, and the field electrode, respectively. The axis labels are local and do not correspond to CMS coordinates. Reprinted from [4]

An individual CSC module has a trapezoidal shape. The trapezoids are arranged into annuli of multiple size. There are four muon station layers in the z-direction for each endcap. In the layer closest to the interaction point, there are three CSC annuli, with smaller ones fitting into the inner circles of the bigger ones. The second and third layers have two annuli, while the fourth layer has only one close to the beamline. There are 36 or 18 partially overlapping trapezoids per annulus.

Internally, each CSC module consists of seven trapezoidal panels stacked with 9.5 mm gas gaps in between. The second, fourth, and sixth panels have anode wires wound around and are suspended at 4.75 mm from the panel surface. The anode wires are gold-plated tungsten with 50 μm diameter, and are separated by 3.2 mm from each other. Six of the panels have cathode strips on one side, thus forming six independent chambers per module. The strips have 8 mm to 16 mm pitch depending on r. The dimensions are somewhat different for the chambers in the innermost stations due to less available space and required higher precision. Those chambers are also placed within the 3.8 T magnetic field, and thus have their anode wires slightly tilted to counteract the Lorentz drift of the electrons. The gas used in the CSC is a 40/50/10 mixture of Ar, CO_2, and CF_4, respectively.

As already mentioned, the charge signal induced by the avalanche ionization in the cathode strips is used for a precise position measurement. The signal amplitude from each strip is digitized by a 12-bit ADC. Since a single charged particle in general leaves a signal in multiple strips per gas gap, by taking the center of mass of the cathode charges in each gap, a spacial resolution of 150 μm in the azimuthal direction is achievable. The anode data, on the other hand, is binary, and records only whether a hit was registered for each bunch crossing.

3.2.4.3 Resistive Plate Chambers

A resistive plate chamber is a parallel-plate gaseous detector with a nanosecond-level time resolution. RPCs are employed in CMS to complement the wire chambers, since the time between the particle arrival and detection in the latter is often longer than 25 ns, leading to a possible ambiguity in the bunch crossing assignment to a given muon signal. In particular, a single drift tube cannot distinguish between ionization occurring at points separated by a distance corresponding to a 25 ns drift. The combination of signals from multiple DT cells and the use of pattern recognition can resolve this ambiguity most of the time, but such signal processing might not be fast enough for triggering at high rates. RPCs, with their superior timing resolution, can unambiguously assign a bunch crossing to each muon signal without sophisticated algorithms and are thus used in the CMS trigger system.

The fast response of the RPC is due to its parallel-plate geometry. Parallel-plate detectors such as spark chambers and RPCs work by applying a very high voltage between two large plates separated by a gas gap. Ionization caused by a passing charged particle in the gas gap is amplified immediately without drifting parallel to the plates. Spark chambers use conducting plates and are operated at above the breakdown voltage, i.e., a conducting filament of plasma is formed at the ionization point and the plates keep discharging unless the high voltage is cut. RPCs, on the other hand, use high-resistivity electrodes for the parallel plates. Therefore, a continuous discharge of the plates does not occur in RPCs, and the ionization avalanches are completed when all of the ionized charge reaches the electrodes. The charge deposited on the surface of the resistive plates dissipates with a relaxation time of $\mathcal{O}(10\text{--}1000)$ ms [10], which becomes a deadtime for the affected region. However, when operated below the breakdown voltage, reionization due to the de-exitation photons from the gas molecules is negligible, and therefore the avalanches are strongly localized. Thus the deadtime is negligible unless under extremely high flux. The local ionization avalanche is read out as a signal using the image current it induces on independent readout strips separated from the electrodes.

The electrodes of the CMS RPC are made of a plastic material (bakelite). The chambers have a double-gap structure, where a layer with the readout strips is inserted in between two 2 mm-thick bakelite plates. The gap on each side is also 2 mm thick and is filled with a gas whose dominant component is $C_2H_2F_4$. The double-gap structure allows a lower applied voltage between the plates than otherwise and enhances the detection efficiency. RPCs were operated at approximately 9.5 kV per gap during Run I.

One barrel module of RPC has a length of 2455 mm, which fits in one wheel. The modules are located inside the muon stations with DT chambers. The first two inner stations have two RPC modules, resulting in a total number of six radial RPC layers. The readout strips run along z and have pitches corresponding to $\Delta\phi = 0.3°$. The segmentation in the z-direction is coarser; strips in each module are divided into two or three sections.

The endcap RPC modules have trapezoidal shapes like the CSCs. They are inserted in the three station layers nearest to the interaction point. The readout strips run radially with the same $\Delta\phi$ pitch as in the barrel. In the radial direction, there are six strip segments in total for each layer.

3.3   Trigger System

The purpose of the CMS trigger system is to pick out a few hundred most interesting hard-scattering events from tens of millions of bunch crossings per second.[2] The rate reduction is necessary because it is not possible to record and analyze every collision event that takes place at CMS. In addition, for the purpose of searching for new phenomena at high-energy scales, soft-scattering events that constitute nearly all of the collisions are assumed to be uninteresting. On the other hand, collision data are extremely precious, and therefore the loss of important data must be avoided. The trigger system was thus designed to carry out an online, real-time selection of events of broad interest within the allowed bandwidth.

CMS has a two-level trigger system. The level-1 trigger (L1T) consists of on- and off-detector electronics that perform rough estimations of the event parameters and select events based on loose requirements. The second level, called the high-level trigger (HLT), receives fully built event data from the data acquisition system (DAQ) on bunch crossings accepted by the L1T, and runs software-based analyses of the events on a large "farm" of commercially available PCs. The system design demands that the L1T output rate is less than 100 kHz, which the HLT further reduces to a few hundred Hz.

The trigger rate and correspondingly the breadth of the recorded event types is highly configurable in CMS. The L1T employs field-programmable gate arrays (FPGA) in many parts of its logic, giving flexibility to define the thresholds as seen fit to the running condition. The HLT is even more, in fact virtually completely, modifiable since it is based on software. Thus the L1T output rate of 100 kHz is in principle a limitation on the DAQ and event building. The ultimate bottleneck during Run I was actually the offline prompt processing of data that took place 48 h after the data was taken. The prompt processing could only run in a timely manner for

---

[2] About a third of the LHC buckets are not filled even at the maximum instantaneous luminosity. Therefore, while the design minimum spacing between bunches is 25 ns, the average bunch crossing rate is less than 40 MHz.

event numbers that correspond to approximately 200 Hz of HLT acceptance. Since data recording could run at more than 800 Hz, the last period of Run I produced the so-called parked data, which were recorded but processed only during the long shutdown that followed Run I.

ECAL, HCAL, DT, CSC, and RPC participate in the L1T. The trigger is naturally composed of two "paths": the calorimeter trigger, using inputs from the ECAL and HCAL; and the muon trigger, using the muon trackers. The calorimeter trigger calculates quantities related to electron/photon (e/γ) and jets as well as global variables such as total hadronic activity and E_T^{miss}. Jets, described in detail in Sect. 3.4.5, are signatures of hard-scattering events with hadronic final states. The muon trigger is dedicated to finding muons and estimating their momenta. Information from the two trigger paths is combined in the global trigger (GT) processor, where the final decision on accepting or rejecting the event is made.

The basic input to the calorimeter trigger are the trigger primitives from the ECAL and HCAL. The TP is a coarse but fast form of energy deposit data generated at every bunch crossing. The ECAL creates a TP for each trigger tower, summing the ADC counts of all the crystals in the tower. The width of the electromagnetic shower, if present in the tower, is also assessed by monitoring the localization of the energy deposit within each tower. A tower considered to have a narrow energy deposit, characteristic of an e/γ signal, is given what is called the fine-grain flag. The HCAL TP is generated for each HCAL tower by summing the E_T from 2 or more neighboring time samples.

The TPs from the ECAL and HCAL are sent to the regional calorimeter trigger (RCT). The ECAL and HCAL towers sharing the same η-ϕ coordinate are bundled into one trigger tower in the RCT. The entire η-ϕ space of calorimeters is split into 4×4 tower regions which are processed independently by the RCT. Within each region, the RCT performs the following:

- Identify e/γ candidates and classify them according to the surrounding activity (isolation).
- Tag hadronic clusters possibly originating from decays of τ leptons.
- Calculate the total energy deposit in the region.
- Find energy deposit patterns consistent with a minimum ionizing particle (MIP) such as a muon passing through.

The identification of e/γ candidates and τ tagging use the fine-grain flag and the energy deposit ratio between the ECAL and HCAL.

The calculation results of the RCT are then combined in the global calorimeter trigger (GCT), which processes the entire calorimeter space and completes the full description of the event from the calorimetric perspective. Aside from sorting the e/γ candidates and τ-tagged jets by their E_T and quality, the GCT merges the RCT E_T sums to form generic jet candidates, and from them calculates event-wide variables such as E_T^{miss}. For each event, up to four isolated and four non-isolated e/γ candidates, up to four central, four forward, and four τ-tagged jets, E_T^{miss}, total E_T, total hadronic E_T, and missing hadronic E_T are output from the GCT.

A similar hierarchical processing takes place in the muon trigger. The DT and CSC first perform the reconstruction of track segments local to individual chambers. In the DT, small segments are first identified from the hit patterns in each superlayer, which are subsequently connected with those from other superlayers. The CSC finds patterns in the anode and cathode data independently, and combines both to form three-dimensional segments. These local segments are then passed to what are called track finders, which use the geometrical relations between the muon stations to form track candidates for the DT and CSC. Since the RPC has only one readout layer per station, it does not perform any local segment reconstruction. The coincidence of hits in different stations is directly used to form RPC tracks. In the barrel, data from the HO is combined with RPC hits to reduce the accidental background.

Track candidates from three muon trackers are passed to the global muon trigger (GMT), which merges overlapping tracks and assesses the quality of each track. MIP data from the RCT is fed to the GMT through the GCT. The GMT also receives isolation information from the GCT and includes it in the quality calculation. The GMT output for each event contains up to four muons with the quality data attached.

The GT receives input from the GCT and the GMT. At this point, the raw detector responses are superseded by physics objects, i.e., e/γ and μ candidates, τ-tagged and generic jets, energy sums, and energy deficits. Thus the responsibility of the GT is to apply kinematic and topological selections to the events so that interesting events get accepted to the HLT at a rate below 100 kHz. An example of a GT selection is the muon plus e/γ trigger described in Sect. 4.2, which requires at least one muon with $p_T > 3.5$ GeV and at least one e/γ object, either isolated or non-isolated, with $E_T > 12$ GeV. Arbitrary combinations of objects and thresholds are allowed. There are 128 programmable selections, called algorithmic bits, in the GT. The full set of selections is called the GT menu, which was revised multiple times during Run I to reflect the changes in the LHC beam energy and intensity. On a L1 accept, a 128-bit string indicating which of the algorithmic bits fired is sent to the HLT along with the trigger objects that caused the bits to fire.

The full detector readout from each bunch crossing is pipelined, or held in a buffer memory, locally at each sub-detector. When the L1 accept decision for a bunch crossing is returned to the sub-detectors, they accordingly send the data to the DAQ. Data for bunch crossings that are not accepted are thrown away. The sub-detector data fragments are sorted and merged in the DAQ into one event data, which is then sent to the HLT farm for further processing.

The HLT, which comprises approximately 13,000 CPU cores, runs the CMS software framework (CMSSW) used in the offline analysis. This design reduces the effort needed to develop the HLT software and provides good homogeneity between the online and offline event selections. It also implies that in principle everything that is done in the offline analysis can also be done online, barring the use of time-dependent calibration constants which are calculated from the recorded data. However, in practice, a severe limitation on the available CPU time requires to abbreviate certain aspects of the event processing such as high-precision tracking. In Run I, one event was processed by a single CPU core and had a time budget of ~200 ms on average.

Various trigger logics implemented in the HLT are called trigger paths. There are more than 400 paths defined to cover a broad range of physics signatures. A full set of trigger paths is called the HLT menu, and is updated frequently following algorithmic improvements, shifts in physics interests, and changes in the beam condition. Each path is seeded by one or more L1 algorithmic bits and tightens the loose L1 selection with more specific criteria. Thus only a fraction of the full HLT menu is executed for any given event. Additionally, the paths are implemented sequentially, applying selections (HLT filters) incrementally on the objects. The processing of a path is aborted as soon as the event fails one HLT filter in the path to minimize unnecessary computations. Following up on the example of the muon plus e/γ trigger, the HLT calculates refined momenta of the objects in the event firing the L1 muon plus e/γ bit, applies quality criteria to the objects, and makes the final decision whether or not to record the event. Since the momentum estimation given in the L1T is not precise, the HLT p_T threshold for a physics objects is usually set higher than in the L1T. Again in the case of the muon plus e/γ trigger, both the muon and e/γ object must have $p_T > 22\,\text{GeV}$ on top of passing strict quality criteria.

Each HLT path is assigned to an output data set called primary data set (PD). A PD is a collection of HLT paths firing on similar event signatures to provide a convenient organization of the recorded data. An event that fires a trigger path is always saved in the PD that the path is assigned to. Since it is not uncommon for an event to exhibit multiple different physics signatures, a single event can end up in multiple PDs. Such overlaps must be taken into account in the offline analysis to avoid double-counting of the events.

As already mentioned, the main objective of the trigger system is to separate interesting hard-scattering events from a far larger pool of soft-scattering collisions. The soft scattering is, however, also of interest in some cases, e.g., when studying the PU background. Therefore there is always a trigger path with only very basic requirements, such as the existence of bunches. The data set collected this way is called minimum bias data.

3.4   Object Reconstruction

Readout from all sub-detectors are merged into events by the DAQ and recorded on disks and tapes. This data, however, only reports what the response of each detector channel was at the time of the given bunch crossing. To explore the physics process that took place at the collision, the readout information must be translated into a physical description of the event. In other words, the particles, or more generally physics objects, that emerged from the pp interaction point in the event need to be reconstructed, combining the readout data with the knowledge of the detector geometry, calibrations, and the magnetic field.

An example of such a reconstruction was already given in the previous section; L1T performs a crude reconstruction of e/γ objects, τ and hadronic jets, and muons

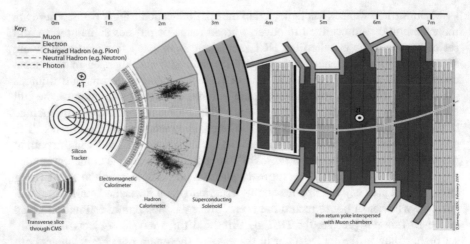

Fig. 3.11 Typical CMS detector response to muons, electrons, neutral hadrons, charged hadrons, and photons. Retrieved from [11]

using fast detector readout. The event reconstruction also takes place in the HLT with higher precision than in L1T. As already mentioned in Sect. 3.3, due to CPU time constraints, the object reconstruction in the HLT is abbreviated and does not bring out the maximum performance of the CMS detector in terms of resolution and efficiency. The best possible reconstruction is executed during the offline analysis, when there is in principle no time constraint.

The object reconstruction makes use of the characteristic signatures that various particles leave in the detector. Figure 3.11 illustrates the overall detector response to different types of particles. As indicated in the drawing, collider-stable particles can be classified by the combination of sub-detector signatures as follows:

- **Muons** leave hits in the inner and outer (muon) trackers, but leave no energy deposit in the calorimeters.
- **Electrons** also leave hits in the inner tracker, but deposit all of their energy in the ECAL and are stopped there.
- **Charged hadrons** have similar signatures to electrons, but their energy deposit in the ECAL is limited, and they are stopped in the HCAL.
- **Neutral hadrons**, on the other hand, leave no trace until they undergo nuclear interactions in the HCAL.
- **Photons** are detected only in the ECAL, where they lose all of their energy via electromagnetic showers.

It should be noted that the above description is for illustration of the ideal case. In reality, muons and electrons can radiate, charged and neutral hadrons can undergo nuclear interaction before arriving at the HCAL, and photons can convert to e^+e^- in the tracker volume. The rather thick inner detector material in CMS makes such scenarios even more likely. Furthermore, when particles scatter, the association of hits in different sub-detectors can fail, leading to inefficiencies and reconstruction of spurious, or fake, objects. Nevertheless, the basic idea of the reconstruction follows from the five patterns described above. The reconstruction software is written accordingly, with tunes and additions to succeed in the non-ideal cases as often as possible.

In the reconstruction process, the detector signals are first interpreted as hits, or representations of the interactions of a single particle with the detector material. The definition of a hit is dependent on the sub-detector. For example, in the pixel tracker, a hit is a cluster of recorded charge deposit in neighboring pixels, while in DT it is made of the TDC readout of a single cell, and represents a pair of possible ionization centers symmetric about the anode wire.

3.4.1 Tracks

Tracks detected by the inner silicon trackers form the basis for all particle reconstructions. Track reconstruction serves three purposes:

- **Particle identification.** Presence or absence of an inner track associated with hits in the calorimeters or the muon system is an important clue on which particle caused the outer hits.
- **Momentum and charge measurement.** Through Eq. (3.1), the curvature of the trajectory is translated to the p_T of the particle under the uniform 3.8 T solenoidal field. The z-component of the momentum is inferred from the inclination of the trajectory helix in the r–z plane. The direction of the bend in the transverse plane depends solely on the sign of the particle charge.
- **Vertex reconstruction.** When a sufficient number of charged particles emerge from a pp collision, the tracks reconstructed from their traces gather around a single point when extrapolated back to the interaction region. The interaction vertex is reconstructed at the point where the likelihood of all such trajectories passing it is the maximum. Typically there are multiple such vertices in an event, corresponding to the PU interactions. The one with the maximum sum of the square of the p_T of all the associated tracks ($\sum p_T^2$) is called the primary vertex and usually corresponds to the hard-scattering interaction.

CMS tracking starts from finding trajectory seeds, or small sets of tracker hits whose patterns are consistent with particle trajectories. This step is performed in iterations, with the searched patterns becoming less trivial in each iteration. Hits used in trajectory seeds in each iteration are masked and excluded from later iterations, so that the more difficult pattern recognition schemes are performed on a smaller number of hits. Once all possible seeds are identified, each trajectory is propagated outwards and inwards to find other tracker hits compatible with it. Since the parameters of the trajectories such as the curvature evolve as the particles scatter or lose energy through ionization, a Kalman filter [12] technique is used to adaptively estimate the parameters during the search for hits. A final fit of the trajectory shape is performed on each fully reconstructed track to refine its position and momentum measurements. The χ^2-value of the fit is saved together with the track parameters.

The number of reconstructed tracks per event depends largely on the event selection. Minimum bias events in the last period of Run I typically featured ~800 reconstructed tracks, most of which had $p_T < 2\,\text{GeV}$. Each track had 3–30 hits in the inner tracker. The tracks with more hits than the number of tracker layers in CMS are the so-called loopers, or tracks with p_T too low to reach the EB and thus are returned to the tracker volume by the magnetic field.

3.4.2 Photons

The core of a photon object is a cluster of ECAL energy deposit. In most of the events there are numerous clusters, since light neutral mesons such as π^0 decay to photons. Additionally, electrons also cause electromagnetic showers in the ECAL. Usually these clusters are not of interest for an analysis using photons as the physics signature. However, it is left to the downstream analysis to distinguish prompt photons, i.e., those that originate in electroweak hard scattering, from hadron decay products and electrons. At the reconstruction stage, all reconstructed ECAL clusters that pass a very loose preselection are labeled as photon objects.

The ECAL energy deposit is first calculated for each crystal using the output pulse of the ECAL MGPA sampled at 40 MHz. Ten time samples around the triggering bunch crossing time stamp are recorded for each crystal. The pulse amplitude and shift in time can be recovered from these ten samples. Together with calibration constants, the amplitudes are converted into the energy and time of the shower signal as seen by the crystal. Pulses that are in-time with the triggering bunch crossing are clustered together according to geometrical proximity. Since a significant portion of photons convert, and electrons radiate, a single physical photon can often leave multiple small clusters in the ECAL. Such small clusters are called basic clusters. Thus the final objects to be identified as photons are clusters of basic clusters, called superclusters, which have a wider extent in the ϕ-direction than in the η-direction because electrons both in the tracker volume and in the electromagnetic showers are diverted in ϕ due to the magnetic field.

The first step of clustering is the identification of the seed crystal, which is a crystal with energy above a given threshold that is higher than any of its neighbors. The neighboring crystals are connected to the seed, expanding outward. Different algorithms are employed in EB and EE. In EB, clustering starts with a 5×1 strip of crystals in $\eta \times \phi$ centered at the seed crystal. The strips of the same orientation are added in the ϕ-direction for up to 17 rows in both the positive and the negative ϕ-direction. Multiple basic clusters can be present within the covered region. The collection of basic clusters found in the region constitutes the supercluster. The clustering in the EE follows a more bottom-up approach and starts with identifying 5×5 crystal matrices, which are the basic clusters, around seed crystals. If two basic clusters are close enough in η and ϕ, they are merged into a supercluster. Hits in the preshower detector are linked to the resulting superclusters.

From the resulting collection of superclusters, the ones with $E_T > 10\,\text{GeV}$ are used for photon objects. If E_T is less than $100\,\text{GeV}$, the clusters must additionally satisfy $H/E < 0.5$, where H/E is the ratio of the HCAL energy deposit found behind the supercluster to the energy of the supercluster. The momentum vector assigned to each photon points from the primary vertex to the center of the supercluster. The center of the supercluster is given as the energy-weighted mean of the basic cluster positions, each of which, in turn, is calculated as an energy-weighted mean of the crystal positions, with parametrized shower depth taken into account.

Supercluster energy is used as the magnitude of the momentum vector of the photon after certain corrections. First, energy lost in the ECAL dead region is estimated and added back to the supercluster. In addition, the energy is scaled by a correction function dependent on the pseudorapidity, shower profile, and E_T of the supercluster. After these corrections, the energy resolution of the photons is shown to be better than 3% in the EB and 5% in the EE [13].

3.4.3 Electrons

An electron object is essentially an ECAL cluster with an associated track. The ECAL cluster is the supercluster mentioned above, which is in fact taken from the identical cluster collection used in the photon reconstruction. The tracks for electrons are, however, constructed differently from the generic tracks, since electrons have higher probability of scattering in the tracker volume than other charged particles. To accommodate sudden changes in the trajectory, which can be drastic in some cases, a Gaussian sum filter (GSF) tracking [14, 15] is performed on track candidates for electron reconstruction.

The association between superclusters and GSF tracks is done in two directions: outside-in and inside-out. The outside-in method is called ECAL-driven electron seeding, and proceeds by iterating over all superclusters selecting trajectory seeds that are compatible with the assumption that the supercluster is due to an electron or a positron. Each selected trajectory seed is projected outwards through GSF tracking, and the resulting track is once again tested for geometrical compatibility with the supercluster. Only the combinations passing the test are saved as electron candidates. In the inside-out method, called tracker-driven electron seeding, all fully reconstructed generic tracks are considered. Each track trajectory is extrapolated to the ECAL surface to find a matching cluster. Those that have a match or pass certain quality criteria are re-reconstructed through the GSF algorithm and then passed through a selection based on a boosted decision tree (BDT). Small ECAL clusters that occur due to bremsstrahlung photons are added to the electron object during the GSF track reconstruction.

The tracker-driven seeding is more efficient for low-p_T electrons than the ECAL-driven method. For isolated electrons with moderate to high p_T, the gain in the reconstruction efficiency from having redundant methods is small; for $p_T > 25\,\text{GeV}$,

ECAL-driven reconstruction captures 92% of all electrons, and the tracker-driven reconstruction adds less than 3% on top of that [15].

The momentum assignment to each electron object depends on the classification of the object based on the fraction of energy lost to bremsstrahlung and the quality of track reconstruction. The ECAL energy measurement and the tracker momentum measurement are combined in such a way to minimize the momentum uncertainty for each electron. A similar supercluster energy correction as the one applied to photons is also applied to electrons. For a "golden" electron, whose track is in the highest quality class and the shower is contained in a narrow cluster, its momentum resolution is better than 2% in the barrel [15].

3.4.4   Muons

A muon in CMS can be identified, or tagged, as a standalone muon, a tracker muon, or a global muon. The tags are not mutually exclusive. A standalone muon is based on a track reconstructed in the muon trackers. The standalone reconstruction starts with track segments in the muon stations. The segments are then extended by the same tracking algorithms employed in the inner trackers. A tracker muon is the muon-equivalent of a tracker-driven electron; each general track with $p_T > 0.5$ GeV and $p > 2.5$ GeV is tested for a corresponding muon segment and, when such segment is found, saved as a tracker muon. A global muon is reconstructed from standalone muons by searching for compatible inner tracks. A new track fit is performed for the combined inner and outer tracks for each global muon to obtain the best measurement of the muon momentum.

The reconstruction efficiency of muons with $p_T > 25$ GeV is greater than 99% in the barrel and slightly lower in the endcap. The overall momentum resolution is 1.8% [6].

3.4.5   Jets

A jet is a collimated flux of hadrons and hadron decay products which signifies a high-energy hadronic interaction. In the perturbative picture of QCD, which is accurate at high-energy scales (see Sect. 2.3), the core of a pp collision is a scattering of quarks and gluons, often called partons, into final-state elementary particles, which can also be quarks and gluons. The partons in the final state, however, cannot be described by perturbative QCD once they become separated apart, participating in long-range QCD interactions. Due to color confinement (see Sect. 2.4), each parton will hadronize with unit probability. How exactly the hadronization proceeds is, however, undetermined; through non-perturbative QCD interactions, hadrons are stochastically produced around each parton. What is known is that the distribution of the magnitude of the momenta of such hadrons relative to the original parton follows a steeply falling curve. The characteristic spread of this distribution of about $\sim 300\,\text{MeV}$, an energy scale often labeled Λ_{QCD}, is where $\alpha_3 \sim 1$. Thus, for outgoing partons with $p_T \gg \Lambda_{\text{QCD}}$, the "swarm" of low-momentum hadrons around the parton becomes collimated along the parton momentum in the laboratory frame, and form the jet. To reconstruct such jets, the final-state hadrons must be grouped according to some definition of proximity between two objects. The grouping of hadrons into jets is called jet clustering.[3]

CMS employs the anti-k_T algorithm [16] for jet clustering. In this algorithm, each object is assigned a "distance" to the beam

$$d_{iB} = p_{\text{T}_i}^{-2},\tag{3.4}$$

where i is the index of the object. The distance between two objects is given by

$$d_{ij} = \min(p_{\text{T}_i}^{-2}, p_{\text{T}_j}^{-2})\frac{\Delta R_{ij}^2}{R^2},\tag{3.5}$$

for $\Delta R_{ij}^2 = (\eta_i - \eta_j)^2 + (\phi_i - \phi_j)^2$ and an arbitrary parameter R. Clustering proceeds by finding two objects with the smallest distance and merging them into one object, adding their momenta. If the smallest distance is d_{iB}, then the object i is regarded as a jet and is removed from the list. The procedure is repeated until all objects are removed. The resulting clusters are more conical than other common clustering algorithms such as k_T clustering. The distance parameter R roughly sets the cone size. A value of $R = 0.5$ was found to be wide enough to capture the partonic structure of the hard scattering, and at the same time narrow enough to avoid large uncertainties due to the inclusion of PU energy deposits into the jets.

The constituents of jets in CMS are called particle-flow (PF) objects, which is described in detail in Sect. 3.4.6. PF uses the full information available from the CMS detector from each event and creates a nearly complete list of observable particles. By using PF objects, most of the hadrons with $p_T \gtrsim 500\,\text{MeV}$ can be included in jets, making the reconstructed jet energy close to its true value. Consequently the calibration scale (jet energy scale, JES) applied to the jets can be close to unity, resulting in a smaller uncertainty.

---

[3]For a reason discussed in Sect. 3.5.1, there is a certain arbitrariness in the distinction between perturbative hard partons and soft QCD emissions. Accordingly, this description of jet is somewhat ill-defined. In fact, there is no self-evident way how the jets should be reconstructed. Nevertheless, the observable fact is that the hadrons in each event can be clustered in one way or another, and the properties of the clusters show correlations to other physical quantities. This situation suggests a more pragmatic approach, which consists of agreeing on a specific clustering algorithm when performing a theoretical calculation and an experimental analysis. Theoretical predictions are then tested in terms of the clusters, rather than by reduction of the observed objects to partons. By using MC simulations as described in Sect. 3.5 to calculate theoretical predictions, this approach is inherently built into collider physics analyses.

In practice, the jet energy correction comprises a subtraction of the energy offset due to PU and the application of the JES [17]. The uncertainty on the JES is at a few % level for jets with $p_T > 40$ GeV. The resolution of the jet energy itself, i.e., the spread of the discrepancy of the clustered energy from the energy of the original parton, is 15–20% at $p_T = 40$ GeV and goes down to 5% with increasing energy. The resolution gets worse with higher PU.

3.4.6 Particle Flow and Missing Transverse Energy

PF is a reconstruction method with the purpose of providing a global description of each event using all available information from the entire CMS detector. In more concrete terms, PF processes the hits from all sub-detectors and interprets them completely in terms of muons, electrons, charged hadrons, neutral hadrons, and photons. It was already shown above that any object reconstruction will make use of multiple sub-detectors. The difference between PF and a more traditional approach of reconstructing different objects independently is that the use of hit information is unique, i.e., one hit is associated with one and only one particle.

The full event reconstruction allows soft particles that are otherwise ignored to be used in the analysis. Moreover, the identification of different particle types enables the application of particle-specific momentum calibrations. For example, a charged pion and an electron both leave hits in the silicon tracker and deposit energy in the calorimeters, but the ratio of observed to true energy is in general lower for pions.

The biggest beneficiaries of such a finely calibrated global event description are jets, which are discussed in Sect. 3.4.5, and E_T^{miss}. PF also enhances the performance of τ lepton reconstruction. The advantage of PF for accurate E_T^{miss} reconstruction is rather obvious; the parts of an event that are missing will become most clear when the parts that are present are known well. E_T^{miss} is calculated as the inverse of the vector sum of the transverse momenta of all PF particles. A possible bias in E_T^{miss} due to unreconstructed particles is corrected using the JES mentioned in Sect. 3.4.5, by adding to raw E_T^{miss} the differences in p_T vectors of fully corrected jets and offset-subtracted raw jets.

The automatic identification of individual particles is both the strength and a source of potential problems of PF. In general there is no perfect criteria to distinguish a particle type from fake objects, and lines are drawn balancing the identification efficiency and the purity of the selected objects. The required balance depends on the specific usage, which leads to the definitions of "working points" or predefined selection criteria corresponding to various points in the efficiency-purity curve (ROC curve). For muons, the efficiency and purity can be both close to unity, but the same cannot be said for electrons, photons, and hadrons. Therefore, during Run I, if an analysis was critically sensitive to the purity and the efficiency of electrons and photons, those reconstructed through dedicated algorithms described in Sects. 3.4.2–3.4.3 were used. Muon reconstruction, on the other hand, was fully incorporated in particle flow.

3.5 Physics Simulation

Modern collider physics analysis is virtually impossible without simulations of the primary scatterings and the subsequent detector response. Simulations are utilized for many purposes in an analysis, including the calculation of expected observables from known or hypothetical processes, calibration of the detector, and test of analysis methods. Some use is more focused on the detector itself than the physics of the scattering. In fact, the simulation of the detector is usually an independent step from that of the primary scattering, since the latter, called event generation, only depends on the types of the incoming particles and their center-of-mass energy.

3.5.1 Monte-Carlo Event Generators

Monte Carlo (MC) event generators are used to simulate the physics of particle scattering. For a given initial state and a list of final-state particles, points in the final-state momentum phase space are randomly sampled following a probability distribution which is proportional to the differential cross section of the process. The expression Monte Carlo, after the famous Monte Carlo casino house in Monaco, refers to the use of random numbers in the event-generation process. The initial state in the present case is two protons colliding at $\sqrt{s} = 8\,\text{TeV}$.

In the calculation of the differential cross section of a pp collision event, the complications from the non-perturbativity of QCD need to be taken into account. The paradigm for dealing with high-energy hadron collisions is established by the QCD factorization theorem. This theorem states that the probability of obtaining some final state from a hadron collision can be calculated as the product of the probability that specific partons are involved in the interaction and the probability for these partons to produce the desired final state. If the final state involves hadrons, the latter part is further factorized into parton-to-parton scattering and the formation of final-state hadrons from outgoing partons. The master formula for the differential cross section calculation in proton–proton collisions is thus

$$\frac{d\sigma_{\text{pp}\to X}}{d\Phi} = \sum_{\{s_i\}} \int \prod_i \frac{d^3 q_i}{(2\pi)^3 2E_i} \sum_{ab} \int dx_a dx_b (2\pi)^4 \delta^4 \left(x_a p + x_b p' - \textstyle\sum_i q_i\right)$$

$$f_a^{\text{p}}(x_a) f_b^{\text{p}}(x_b) \frac{1}{2\hat{s}} \overline{|\mathcal{M}|^2} (ab \to \{s_i\}; x_a, x_b, \{q_i\}) D\left(\{s_i, q_i\}; X(\Phi)\right). \quad (3.6)$$

This formula spells out the factorization of the cross section calculation. First, partons a and b are selected from the two incoming protons. Parton a carries a fraction x_a of the momentum p of one of the protons, and b a fraction x_b of p', the momentum of the other proton. The number of parton a within a proton with momentum fraction x_a is given by $f_a^{\text{p}}(x_a) dx_a$. The function f_a^{p} is called the parton distribution function (PDF) of the proton for parton a. The factor

$$\frac{1}{2\hat{s}} \overline{|\mathcal{M}|^2} (ab \to \{s_i\}; x_a, x_b, \{q_i\}) (2\pi)^4 \delta^4 \left(x_a p + x_b p' - \textstyle\sum_i q_i\right) \frac{d^3 q_i}{(2\pi)^3 2E_i}, \quad (3.7)$$

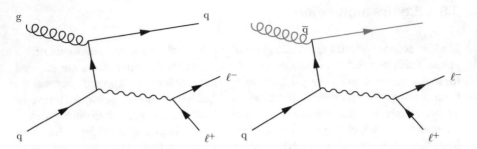

Fig. 3.12 *Left*: gluon-quark scattering with gluon splitting g → q$\overline{q}$. *Right*: quark-antiquark scattering. The two diagrams represent an identical hard-scattering process but with different factorizations

where $\hat{s} = (x_a p + x_b p')^2$, is the differential cross section of the $ab \to \{s_i\}$ partonic process with the momenta $\{q_i\}$. $\mathcal{M}$ is the invariant scattering matrix element, and the bar notation in $\overline{|\mathcal{M}|^2}$ indicates averaging over initial- and final-state quantum numbers that are not observable. The product of the PDFs and the partonic cross section is then summed over the momentum fractions and the parton flavors. The last factor D is the fragmentation function, and encodes the transition probability of the outgoing partons $\{s_i\}$ into the final state X which is dependent on some set of variables Φ.

The factor $\overline{|\mathcal{M}|^2}$ in Eq. (3.7) encodes the parton hard scattering. The factorization theorem guarantees that, under certain conditions, this factor is calculable perturbatively. The non-perturbative effects are contained in the PDF and the fragmentation function. Note, however, that the factorization is not unique. For example, it is easy to see that an interaction that is regarded as g q scattering where the gluon splits into a q$\overline{q}$ pair can also be calculated as $\overline{q}$q scattering, if the quark from the gluon splitting is considered nearly collinear with the beam. The two points of view are shown in Fig. 3.12. The classification of the initial state in hadronic interactions is thus somewhat arbitrary, requiring a specification of how the calculation is factorized. Conventionally, the latter is achieved by an energy scale called factorization scale μ_F, which can be regarded as the cutoff to separate perturbative and collinear regimes [18–20].

Event generation also follows this factorized approach. Taking advantage of the fact that PDF and fragmentation functions are independent of the partonic process, the three calculations are often performed in different programs or different sections of a single program.

The PDF currently cannot be calculated from first principles, and is therefore obtained through parametrizations of experimental data. There are multiple publically available PDF sets, such as those from the CTEQ [21, 22] and NNPDF [23] collaborations. The MC simulation samples described in this thesis all employ the CTEQ6L1 PDF set. Software functions for the PDF calculation take the x-value and μ_F as arguments and return the PDF value calculated using the input data set.

The fragmentation function is also in the realm of non-perturbative QCD and thus is currently not calculable. Additionally, due to the stochastic nature of hadronization, there are intractably many combinations of final-state hadrons for a given outgoing parton. Therefore, for hadronic processes, it is not practical to actually specify the true final state and calculate the formula in Eq. (3.6). Instead, partons are given as the final states, and their subsequent soft-parton emissions and hadronizations are obtained by what are called parton-shower MC simulations. The parton-shower simulation assigns each parton a virtuality value, which can be regarded as a sort of energy budget, and randomly makes each parton split, or fragment, into two. The products of fragmentation will have virtuality values lower than their mother. The process is repeated until all partons have their virtuality below a threshold, at which point they are hadronized. The probability of each fragmentation and hadronization history can be calculated using splitting kernels, which are related to fragmentation functions and are given by parametrizing the experimental data. This probability is multiplied to the parton differential cross section to obtain the full weight of the event. Usually, the parton-shower simulation also generates the so-called initial-state radiations (ISR) using a similar algorithm. ISR is the emission of partons from the "remainder of the protons"; through using a PDF to pick out the partons that participate in the hard scattering, the recoil of the other parts of the incoming protons are approximated to be collinear to the beam. In reality, there are finite recoils with a steeply falling p_T distribution, which constitutes ISR.

There are numerous software packages that are capable of parton-shower simulation, such as PYTHIA [24, 25], SHERPA [26], and HERWIG [27, 28]. PYTHIA 6.4 was used for all simulations in this thesis. All of the three programs in fact include the calculators of PDF and the partonic cross sections in Eq. (3.6). Therefore, it is possible to perform the entire event generation with these programs. For this reason, they are called general-purpose event generators.

Unlike PDF and fragmentation function, the invariant matrix element is specific to individual processes. In general, each process has to have its own implementation. Accordingly, most of the general-purpose event generators, including PYTHIA and HERWIG, have a fixed library of final states. There are, however, programs called matrix-element generators, such as MADGRAPH 5 [29], which can automatically write the source code for arbitrary physics processes and execute it. With a standardized interface data format [30], the partonic output of the matrix-element generators can be passed to general-purpose event generators for parton shower.

Besides supplementing the physics libraries of the general-purpose generators, the matrix-element generators have an important usage, which is to generate processes with extra emissions of quarks and gluons in addition to the primary process of interest. The additional partons replace the ISR provided by the parton-shower simulation in the high-p_T regime. This prescription is necessary for high-energy primary processes such as heavy resonance production. When the invariant matrix element only accounts for such processes, the factorization scale must inevitably be set high, and phenomena that are thus considered as uninteresting collinear physics become actually non-negligible effects, such as high-p_T jets. In other words, the assumption of collinearity is broken for high-energy processes. The situation is saved by properly re-factorizing the perturbative and non-perturbative physics, i.e., by generating hard emissions together with the primary process.

Before concluding this subsection, it must be noted that the total cross section for a given final state is calculated whenever a MC event generator is used. This point can be understood by invoking the general relation regarding the multi-dimensional integral over some volume Ω:

$$\int_{\Omega} d\mathbf{x} f(\mathbf{x}) = \Omega \lim_{N \to \infty} \frac{1}{N} \sum_{i=1}^{N} f(\mathbf{x}_i) \tag{3.8}$$

for $\mathbf{x}_i$ uniformly distributed within Ω. The calculation of the total cross section is also a multi-dimensional integral in the momentum space of the final-state particles. The points $\mathbf{x}_i$ in this context correspond to the phase space points mentioned above. Thus, by taking a sum of the differential cross sections calculated at a large number of randomly sampled points, the total cross section can be inferred with an accuracy that scales with the inverse square-root of the number of generated points.

Historically speaking, MC methods were first used as an integration technique, and event generation was a by-product. Even today, there are in fact dedicated cross section calculators that do not generate events. Particularly, those that calculate the invariant matrix element at the NLO in QCD or the ones that employ specific CPU speed optimizations tend to focus on integration. Among such integrators are MCFM [31], which calculates cross sections for rare processes in the SM, and PROSPINO2 [32], which calculates sparticle pair-production cross sections.

3.5.2 CMS Detector Simulation

The final-state 4-momentum vectors from the MC event generators are passed to the detector simulation to predict the observable signals. The simulation is capable of tracing particle trajectories through the detector spacial volumes. The collision point, from which the generated particles emerge, can be placed anywhere. Usually a point randomly displaced from the detector origin with a spread of a few cm is chosen to simulate the finite size of the LHC beam.

There are in general two ways to simulate the detector response. One is to fully rely on first principles and trace every particle, including the secondaries that are produced in the electromagnetic showers and nuclear interactions. In this mode of simulation, the detector hits are registered when the particles deposit observable energy to the detector material according to their specific interaction probability. The second mode is to parametrize showers and energy deposits using empirical formulae. The second mode involves much less particle-tracing than the first, and therefore is less CPU-intensive.

CMS implements both types of simulations. The first, based on complete particle tracing, is called full simulation, and the second with parametrizations is called fast simulation. The core software in both simulations is GEANT4 [33]. The central task of the GEANT4 program is the simulation of particles passing through matter. All particles are treated on equal footing, and are traced down until stopped. New particles can be created along the way through interactions with matter. The material and geometry of the objects that the particles pass through can be specified in detail. In the CMS full simulation, the geometry of the detector components is described as precisely as possible, including their mis-alignment. Inactive and active components are taken into account often with their detailed material composition.

The fast simulation, on the other hand, uses a simplified geometry. The silicon trackers are approximated by infinitesimally thin layers, and the calorimeters are made hollow. The energy loss of particles in the tracker layers are accounted for by parametrization. The showerings in the calorimeters are outsourced to a program called GFLASH [34], which utilizes parametrized particle shower libraries. This way, the fast simulation retains the ability to trace the particles from the primary interactions, while bypassing the time-consuming parts of the GEANT4 simulation.

The simulated detector hits are converted to digital readout of the detectors and passed to the standard reconstruction algorithms as if they are real collision data. The full simulation and real collision data are processed through the identical reconstruction. The fast simulation takes some shortcuts also in the reconstruction step. Tracking, which is the single most time-consuming reconstruction process, is expedited by using the so-called MC truth information, or information from the particle simulation level, instead of relying only on the detector readouts. Although special attention is paid to not introduce bias in the reconstruction efficiency, the fast simulation lacks the simulation of specific hardware-originated effects in the tracking for this reason.

Thus, there is in general a tradeoff between speed and accuracy when selecting the simulation method in CMS. The full simulation is used where appropriate, especially for purposes like the detector calibration. The fast simulation is ideal for the production of the signal data set in searches for unknown physics phenomena. In such searches, a large volume of the parameter space must be scanned and thus a large number of events are required, but the signal simulation does not depend highly on an accurate detector modeling.

Because the full simulation and large-scale fast simulations require intensive computing resources, their production is centralized and streamlined in CMS for efficiency. The centralization also avoids configuration mistakes by the individual analysts. Simulation events are organized into data sets according to the generated physics processes and the software version used for their production. Such data sets are referred to as "official" MC data sets, and those which did not go through the central production machinery are often called "private" data sets.

References

1. CERN: CERN Document Server. http://cds.cern.ch
2. LHC Higgs Cross Section Working Group Collaboration: Handbook of LHC Higgs cross sections: 1. Inclusive observables. Technical report. CERN-2011-002 (2011). doi:10.5170/CERN-2011-002. arXiv:1101.0593 [hep-ph]
3. Public CMS Luminosity Information. https://twiki.cern.ch/twiki/bin/viewauth/CMS/LumiPublicResults
4. CMS Collaboration: The CMS experiment at the CERN LHC. J. Instrum. **3**, S08004 (2008). doi:10.1088/1748-0221/3/08/S08004
5. CMS Collaboration: Precise mapping of the magnetic field in the CMS barrel yoke using cosmic rays. J. Instrum. **5**, T03021 (2010). CMS-CFT-09-015. doi:10.1088/1748-0221/5/03/T03021. arXiv:0910.5530 [physics.ins-det]
6. CMS Collaboration: Performance of CMS muon reconstruction in pp collision events at $\sqrt{s} = 7$ TeV. J. Instrum. **7**, P10002 (2012). CMS-MUO-10-004, CERN-PH-EP-2012-173. doi:10.1088/1748-0221/7/10/P10002. arXiv:1206.4071 [physics.ins-det]
7. Particle Data Group, Beringer, J., et al.: Review of particle physics. Phys. Rev. D **86**, 010001 (2012). doi:10.1103/PhysRevD.86.010001
8. CMS Collaboration: 2012 ECAL detector performance plots. Technical report. CMS-DP-2013-007, CERN-CMS-DP-2013-007 (2013)
9. CMS Collaboration: Energy calibration and resolution of the CMS electro-magnetic calorimeter in pp collisions at $\sqrt{s} = 7$ TeV. J. Instrum. **8**, P09009 (2013). CMS-EGM-11-001, CERN-PH-EP-2013-097. doi:10.1088/1748-0221/8/09/P09009. arXiv:1306.2016 [hep-ex]
10. Lippmann, C.: Detector physics of resistive plate chambers. Dissertation, Goethe University Frankfurt (2003)
11. Interactive Slice of the CMS detector (2010). https://cms-docdb.cern.ch/cgi-bin/PublicDocDB/ShowDocument?docid=4172 (visited on 01/26/2015)
12. Frühwirth, R.: Application of Kalman filtering to track and vertex fitting. Nucl. Instrum. Meth. A **262**, 444–450 (1987). HEPHY-PUB-87-503. doi:10.1016/0168-9002(87)90887-4
13. CMS Collaboration: Performance of photon reconstruction and identification with the CMS detector in proton–proton collisions at $\sqrt{s} = 8$ TeV. J. Instrum. (2015, submitted). arXiv:1502.02702 [physics.ins-det]
14. Adam, W., Frühwirth, R., Strandlie, A., Todorov, T.: Reconstruction of electrons with the Gaussian sum filter in the CMS tracker at LHC. J. Phys. G **31**(9) (2003). eConf C0303241.CHEP-2003-TULT009, TULT009. doi:10.1088/0954-3899/31/9/N01. arXiv:physics/0306087 [physics]
15. CMS Collaboration: Performance of electron reconstruction and selection with the CMS detector in proton–proton collisions at $\sqrt{s} = 8$ TeV. J. Instrum. (2015, submitted). arXiv:1502.02701 [physics.ins-det]
16. Cacciari, M., Salam, G.P., Soyez, G.: The anti-k_T jet clustering algorithm. J. High Energy Phys. **0804**, 063 (2008). LPTHE-07-03. doi:10.1088/1126-6708/2008/04/063. arXiv:0802.1189 [hep-ph]
17. CMS Collaboration: Jet energy scale and resolution in the CMS experiment in pp collisions at 8 TeV. J. Instrum. **12**, 02014 (2017). doi:10.1088/1748-0221/12/02/P02014. arXiv: 1607.03663 [hep-ex]
18. Collins, J.C., Soper, D.E., Sterman, G.F.: Factorization of hard processes in QCD. Adv. Ser. Direct. High Energy Phys. **5**, 1–91 (1988). ITP-SB-89-31. arXiv:hep-ph/0409313 [hep-ph]
19. Plehn, T.: LHC phenomenology for physics hunters (2008). arXiv: 0810.2281 [hep-ph]
20. Sterman, G.F.: QCD and jets (2004). arXiv:hep-ph/0412013
21. Lai, H.-L., et al.: New parton distributions for collider physics. Phys. Rev. D**82**, 074024 (2010). MSUHEP-100707, SMU-HEP-10-10. doi:10.1103/PhysRevD.82.074024. arXiv:1007.2241 [hep-ph]

22. Pumplin, J., et al.: New generation of parton distributions with uncertainties from global QCD analysis. J. High Energy Phys. **0207**, 012 (2002). MSU-HEP-011101. doi:10.1088/1126-6708/2002/07/012. arXiv:hep-ph/0201195 [hep-ph]

23. Forte, S., Garrido, L., Latorre, J.I., Piccione, A.: Neural network parametrization of deep inelastic structure functions. J. High Energy Phys. **0205**, 062 (2002). GEF-TH-3-02, RM3-TH-02-01. doi:10.1088/1126-6708/2002/05/062. arXiv:hep-ph/0204232 [hep-ph]

24. Sjöstrand, T., Mrenna, S., Skands, P.Z.: PYTHIA 6.4 physics and manual. J. High Energy Phys. **05**, 026 (2006). FERMILAB-PUB-06-052-CD-T, LU-TP-06-13. doi:10.1088/1126-6708/2006/05/026. arXiv:hep-ph/0603175 [hep-ph]

25. Sjöstrand, T., Mrenna, S., Skands, P.Z.: A brief introduction to PYTHIA 8.1. Comput. Phys. Commun. **178**, 852–867 (2008). CERN-LCGAPP-2007-04, LU-TP-07-28, FERMILAB-PUB-07-512-CD-T. doi:10.1016/j.cpc.2008.01.036. arXiv:0710.3820 [hep-ph]

26. Gleisberg, T., et al.: Event generation with SHERPA 1.1. J. High Energy Phys. **0902**, 007 (2009). FERMILAB-PUB-08-477-T, SLAC-PUB-13420, ZU-TH-17-08, DCPT-08-138, IPPP-08-69, EDINBURGH-2008-30, MCNET-08-14. doi:10.1088/1126-6708/2009/02/007. arXiv:0811.4622 [hep-ph]

27. Corcella, G., et al.: HERWIG 6: an event generator for Hadron emission reactions with interfering gluons (including supersymmetric processes). J. High Energy Phys. **0101**, 010 (2001). CAVENDISH-HEP-99-03, CERN-TH-2000-284, RAL-TR-2000-048. doi:10.1088/1126-6708/2001/01/010. arXiv:hep-ph/0011363 [hep-ph]

28. Corcella, G., et al.: HERWIG 6.5 release note (2002). arXiv:hep-ph/0210213

29. Alwall, J., et al.: The automated computation of tree-level and next-to-leading order differential cross sections, and their matching to parton shower simulations. J. High Energy Phys. **1407**, 079 (2014). CERN-PH-TH-2014-064, CP3-14-18, LPN14-066, MCNET-14-09, ZU-TH-14-14. doi:10.1007/JHEP07(2014)079. arXiv:1405.0301 [hep-ph]

30. Alwall, J., et al.: A standard format for Les Houches event files. Comput. Phys. Commun. **176**, 300–304 (2007). FERMILAB-PUB-06-337-T, CERN-LCGAPP-2006-03. doi:10.1016/j.cpc.2006.11.010. arXiv:hep-ph/0609017 [hep-ph]

31. Campbell, J.M., Ellis, R.K., Williams, C.: Vector boson pair production at the LHC. J. High Energy Phys. **1107**, 018 (2011). FERMILAB-PUB-11-182-T. doi:10.1007/JHEP07(2011)018. arXiv:1105.0020 [hep-ph]

32. Beenakker, W., Hopker, R., Spira, M.: PROSPINO: a program for the production of supersymmetric particles in next-to-leading order QCD (1996). arXiv: hep-ph/9611232

33. GEANT4 Collaboration: Geant4—a simulation toolkit. Nucl. Instrum. Meth. **A506**, 250–303 (2003). SLAC-PUB-9350, FERMILAB-PUB-03-339. doi:10.1016/S0168-9002(03)01368-8

34. Grindhammer, G., Peters, S.: The Parameterized simulation of electromagnetic showers in homogeneous and sampling calorimeters (1993). arXiv:hep-ex/0001020

Chapter 4
Data Collection and Event Selection

4.1 Overview of Data Samples

The analyzed event samples are taken from the CMS pp collision data recorded in 2012. The data corresponds to an integrated luminosity of $19.7\,\text{fb}^{-1}$ [1] at $\sqrt{s} = 8\,\text{TeV}$. The analysis is performed in two channels, the $e\gamma$ channel, where events are selected if at least one photon and one electron are present, and the $\mu\gamma$ channel, where a muon instead of an electron is required in addition to the photon. In both channels, an excess beyond the SM prediction of events with large E_T^{miss} is searched for. Physically, a photon plus τ lepton would be considered as a photon plus lepton signal. However, the τ lepton is unstable and decays either hadronically to light mesons and a neutrino or leptonically to an electron or a muon and neutrinos. While it is possible to identify hadronically decaying τ leptons, the identification efficiency is rather low and thus the addition of the $\tau\gamma$ channel would not improve the sensitivity of this search. Events with leptonically decaying τ leptons ($\tau \to \ell\nu_\tau\bar{\nu}_\ell$) are automatically included in the $e\gamma$ and $\mu\gamma$ channel.

For each search channel, events are required to have fired specific triggers to ensure uniformity in the event properties. Accordingly, the events are collected from the primary data sets to which the triggers are assigned. In 2012, the CMS run was split into four periods, delimited by major version changes in the HLT menu. There are four primary data sets per search channel which are named periods A, B, C, and D. Table 4.1 lists the data-taking periods and their integrated luminosities.

Photons and leptons in the analyzed data sets are subjected to stringent selection criteria, sometimes referred to as "cuts," to improve the purity of the objects. The purity is the fraction of reconstructed photons and leptons corresponding to real physical counterparts, instead of some other particles faking the signal.

Events with signs of instrumental problems are removed from the analyzed data sets. In particular, when a problem is localized, it often materializes as large E_T^{miss} due to the artificial imbalance in the detector signal. Thus it is crucial for an analysis

© Springer International Publishing AG 2017
Y. Iiyama, *Search for Supersymmetry in pp Collisions at* $\sqrt{s} = 8$ TeV *with a Photon, Lepton, and Missing Transverse Energy*, Springer Theses,
DOI 10.1007/978-3-319-58661-8_4

Table 4.1 Data-taking periods in terms of dates and accumulated luminosities

Period	Start date	End date	$\int \mathcal{L}dt\,(\mathrm{pb}^{-1})$
A	April 5	May 7	876.2
B	May 10	June 18	4411.7
C	July 2	September 27	7054.7
D	September 28	December 5	7369.0

searching for high-E_T^{miss} events to properly clean the data sets. An event is vetoed if it showed either one of the following behaviors: CSC hits consistent with beam halo; local high-amplitude signal from HCAL, which is suspected to be due to discharge in a HPD; HCAL hit pattern consistent with the calibration laser light illumination; large ECAL energy deposits at the exact places where the readout is dead; high-amplitude signal from one EE region, suspected to be noise due to an unstable high-voltage; and lack of tracks compared to what is expected from the calorimeter signal, due to an inefficiency of the tracking algorithm.

MC simulation data sets are also used extensively in the analysis, either to estimate the SM background, model the SUSY signal, or test the analysis method. For the first two cases, the events in each data set are assigned a weight

$$w = \frac{\sigma \int \mathcal{L}dt}{N} \tag{4.1}$$

where σ is the calculated cross section of the simulated process, N is the number of events in the data set, and $\int \mathcal{L}dt = 19.7\,\mathrm{fb}^{-1}$ is the integrated luminosity of the recorded data.

4.2　Triggers

The HLT used to collect the signal candidate events for the eγ channel is a diphoton trigger called

```
HLT_Photon36_CaloId10_Iso50_Photon22_CaloId10_Iso50.
```

This trigger fires on the presence of

- at least one isolated narrow ECAL cluster with $E_T > 36\,\mathrm{GeV}$ reconstructed around a L1T e/γ object with $E_T > 22\,\mathrm{GeV}$, and
- at least one other isolated narrow ECAL cluster with $E_T > 22\,\mathrm{GeV}$.

The trigger is assigned to the Photon or DoublePhoton data sets, depending on the run period.

A narrow ECAL cluster for this trigger is defined as having $H/E < 0.1$ and $\sigma_{i\eta i\eta} < 0.014$. The latter variable encodes the cluster width along the η-direction,

Fig. 4.1 Illustration of isolation cone sums of radius $\Delta R = 0.3$

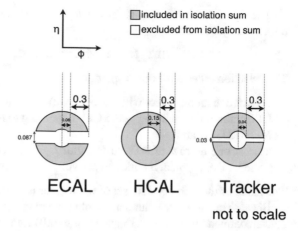

and is defined using the central 5×5 crystal matrix of the supercluster as

$$\sigma_{i\eta i\eta} = (3.045 \times 10^{-4}) \times \frac{\sum_{i \in 5\times5}\left[(4.7 + \log(E_i/E_{5\times5})) \times \Delta\eta_i^2\right]}{\sum_{i \in 5\times5}(4.7 + \log(E_i/E_{5\times5}))}, \tag{4.2}$$

where E_i is the energy of the ith crystal, $\Delta\eta_i$ is the distance of the ith crystal from the center of the matrix in terms of the number of crystals, and $E_{5\times5}$ is the total energy of the matrix. The overall factor of 3.045×10^{-4} is applied for historical reasons.

The isolation in the diphoton trigger is defined by

- $I_{\text{ECAL}} < 5\,\text{GeV} + 0.012E_\text{T}$
- $I_{\text{HCAL}} < 5\,\text{GeV} + 0.005E_\text{T}$
- $I_{\text{Trk}} < 5\,\text{GeV} + 0.002E_\text{T}$,

where I_{ECAL}, I_{HCAL}, I_{Trk} are the sums of E_T of ECAL hits, E_T of HCAL hits, and p_T of tracks, all within a $\Delta R < 0.3$ cone around the photon. Specific regions of the cones are not used in the calculation of the isolation sums to exclude the energy of the photon itself. Figure 4.1 illustrates the different exclusion regions for the three isolation cones.

The diphoton trigger path is implemented in three stages. In the first stage, ECAL clusters are reconstructed locally around the L1T e/γ object, whose momentum directions are passed from the L1T. The second stage is the identification of the leading ($E_\text{T} > 36\,\text{GeV}$) leg, which must come from one of the clusters above. If a cluster passes all selection requirements except for I_{Trk}, the trigger proceeds to the third stage, where all ECAL clusters are reconstructed. The trailing leg can be any of the clusters, and once a trailing cluster that also passes all requirements except for I_{Trk} is identified, track reconstruction takes place to be used for the calculation of I_{Trk}. The final HLT filter in the path checks for the existence of at least two clusters that pass the I_{Trk} requirement. The information on the clusters that pass the leading leg filter and the final filter are saved in the offline data, together with the seeding L1T e/γ objects. The details of the trigger implementation evolved slightly throughout the 2012 run, but the logic and the parameter values were not modified.

The signal candidate events for the $\mu\gamma$ channel are required to fire a muon plus photon trigger called

HLT_Mu22_Photon22_CaloIdL.

The selection criteria for this trigger are

- at least one muon object with $p_T > 22\,\text{GeV}$ and $|\eta| < 2.5$ reconstructed around a L1T muon object with $p_T > 3.5\,\text{GeV}$ appearing together with an e/γ object with $E_T > 12\,\text{GeV}$, and
- at least one narrow ECAL cluster with $E_T > 22\,\text{GeV}$ reconstructed around the same L1T e/γ object mentioned above.

The name of the primary data set of the trigger is MuEG.

Because events with a muon and a photon are rare, the requirements on the ECAL cluster, defined by $H/E < 0.15$ and $\sigma_{i\eta i\eta} < 0.014$, in this trigger is looser than those in the diphoton trigger.

The implementation of the muon plus photon HLT path is also sequential, performing only necessary calculations at any point. Most of the logic is already given as an example in Sect. 3.3. The path is seeded by a L1T muon plus e/γ bit. The HLT receives the momenta of the muons and e/γ objects that fired the seed bit from the L1T. Both muons and ECAL clusters are locally reconstructed around the respective L1T objects. The muons are processed first. The L1T seed objects and the selected photons and muons are saved in the offline data. Again, the details of the implementation evolved, but the logic and the parameter values of this trigger were kept constant over the 2012 run.

Events that fired several other triggers are used in this analysis to perform various background studies and efficiency measurements. The names of their associated data sets and short descriptions of the triggers are listed in Table 4.2.

The two signal triggers are simulated in all of the MC simulation samples that are used in the analysis (see Sect. 4.8.1). However, due to imperfect modeling of the detector response, the efficiency of the triggers to capture events of interest differs slightly between simulation and real data. Therefore, the trigger efficiencies are carefully measured in real data (see Sect. 5.6.3) and are compared to those in simulation to correct this possible bias in the event yield estimation.

Table 4.2 List of auxiliary data sets and the triggers within

Data Set name	Trigger name and description
SingleElectron	HLT_Ele27_WP80
	One isolated electron with $p_T > 27\,\text{GeV}$ and $|\eta| < 2.5$
SingleMu	HLT_IsoMu24 (HLT_IsoMu24_eta2p1 for period A)
	One isolated muon with $p_T > 24\,\text{GeV}$ and $|\eta| < 2.5$ (2.1 for period A)
DoubleMu	HLT_Mu17_Mu8
	Two muons with $p_T > 17\,\text{GeV}$ and $8\,\text{GeV}$

Samples collected with these triggers are used for background studies and efficiency measurements

4.3 Photon Selection

Only photons with $E_T > 40\,\mathrm{GeV}$ reconstructed in the ECAL barrel are used in this analysis. The barrel region is defined by $|\eta_{SC}| < 1.4442$, where η_{SC} is the pseudorapidity of the center of the supercluster with respect to the CMS origin. The EB in fact spans $|\eta| < 1.479$ as mentioned in Sect. 3.2.2, but the clusters at the edge might have part of their energy not reconstructed and thus are disregarded.

The photon object is first required to match the respective trigger objects of the $e\gamma$ and $\mu\gamma$ channels. This trigger-matching assures that the event is indeed collected because of the candidate objects, and that the trigger efficiency measured in Sect. 5.6.3 corresponds to the true efficiency. The match is defined by $\Delta R < 0.3$ for L1T objects and $\Delta R < 0.15$ for HLT objects, where ΔR is calculated using the supercluster position and the momentum vector of the trigger object. Figure 4.2 shows typical ΔR distributions between the online (trigger) and offline objects.

As already mentioned in Sect. 3.4.2, a reconstructed photon object must be further subjected to selection criteria to distinguish prompt photons from electrons and hadron decay products. For each trigger-matching photon object, the following selection criteria are applied in both search channels:

- Selection criteria to reduce hadronic fakes

  - $H/E < 0.05$
  - $\sigma_{i\eta i\eta} < 0.012$
  - $I_{CH} < 2.6\,\mathrm{GeV}$:
    Charged hadron isolation (I_{CH}) is calculated as the p_T sum of the PF charged hadrons satisfying $|d_0| < 0.1\,\mathrm{cm}$ and $|d_z| < 0.2\,\mathrm{cm}$. The impact parameters d_0 and d_z are the transverse and longitudinal distance from the primary vertex to the point of closest approach of the track. The p_T sum is taken within a $0.02 < \Delta R < 0.3$ hollow cone around the photon, similar to the cone for I_{Trk} shown in Fig. 4.1 but without the slit along the ϕ-direction. An estimated offset due to PU is subtracted from the p_T sum. The estimation method is described below. The upper cut of 2.6 GeV is applied after the offset subtraction.
  - $I_{NH} < 3.5\,\mathrm{GeV} + 0.04E_T$:
    Neutral hadron isolation (I_{NH}) is calculated as the p_T sum of the PF neutral hadrons within a $\Delta R < 0.3$ solid cone around the photon. Offset from PU is subtracted. To account for the possible small leakage of photon energy into the HCAL, the upper cut for the p_T sum scales with the E_T of the photon.
  - $I_{Ph} < 1.3\,\mathrm{GeV} + 0.005E_T$:
    Photon isolation (I_{Ph}) is the E_T sum of the PF photons within a $\Delta R < 0.3$ cone around the photon. There is a complication in the calculation because the photon object is reconstructed independently from the PF algorithm. Thus a PF photon must be ignored from the sum if it overlaps with the supercluster of the photon object. Additionally, to avoid counting conversions and bremsstrahlung in the p_T sum, all PF photons that have $|\Delta\eta| < 0.015$ or $|\Delta\phi| < 1$ with respect to the photon object are also ignored. The resulting

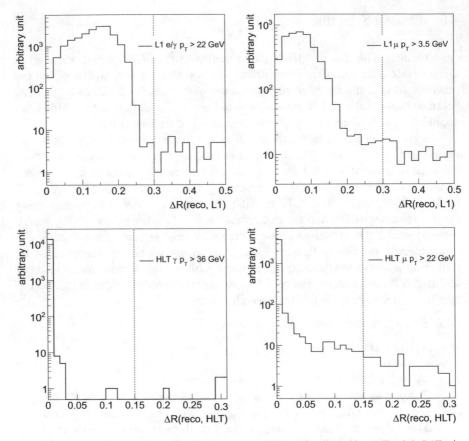

Fig. 4.2 Real-data distributions of the ΔR between offline and online objects. *Top left*: L1T e/γ cluster and offline supercluster. *Top right*: L1T muon and offline muon. *Bottom left*: HLT e/γ cluster and offline supercluster. *Bottom right*: HLT muon and offline muon. The *red lines* represent the selection cuts defining the matches between the two objects

isolation cone is similar to the one for I_{ECAL} in Fig. 4.1. The scaling upper limit is to account for residual leakage of the photon energy into the isolation cone. The PU offset correction is also applied for I_{Ph}.

- Selection criteria to reduce electronic fakes (electron vetoes)

  - Pixel veto:
    No electron track seeds can be associated with the supercluster.
  - GSF veto:
    No reconstructed GSF electrons with $p_T > 2\,\text{GeV}$ can be within $\Delta R < 0.02$ of the photon candidate, where ΔR is calculated using the supercluster positions of the two objects.
  - PF charged hadron veto:

No PF charged hadron object with $p_T > 3\,\text{GeV}$ can be within $|\Delta\eta| < 0.005$ and $|\Delta\phi| < 0.02$ of the photon candidate. The angular variables are calculated using the reconstructed momentum of the PF particles and the vector that points from the vertex of the PF particles to the supercluster positions of the photons.

The PU offset correction for the isolation p_T sums takes two inputs. The first input is the average p_T sum of the objects in the event per area, conventionally denoted as ρ, which is calculated through a dedicated jet clustering with k_T algorithm of all PF particles. The k_T algorithm follows the identical procedure as the anti-k_T algorithm, but the definition of the distance between the objects, which are given by Eqs. (3.4) and (3.5) for anti-k_T, uses p_T^2 instead of p_T^{-2}. The distance parameter of $R = 0.6$ is used in this calculation. For each k_T-clustered jet, the "active area" A [2] is calculated. The value ρ is defined as the median of p_T/A of the jets.

The second input of the PU offset correction is the effective area associated with each photon object. The effective area A_{eff} is calculated from a formula

$$\langle I \rangle = I_0 + A_{\text{eff}}\rho \tag{4.3}$$

which describes the behavior of all the three isolation sums well. Here $\langle I \rangle$ corresponds to the mean of the respective isolation sum, and I_0 is an offset value. The effective area values used for each isolation calculation are summarized in Table 4.3.

Figures 4.3 and 4.4 show the distributions of the variables used for photon selection, first for all photons passing the HLT and then after all the selections are applied except the one being plotted ("$N - 1$" plots). In the figure, the isolation variables are corrected for PU and energy leakage from the photon cluster. All photons in period A are used. Distributions from truth-matched photons in a MC simulation sample are overlaid as references. The normalization of the simulation sample is arbitrary.

The electron vetoes are applied to minimize the probability of an electron to fake a photon. Figure 4.5 shows the distributions of the ΔR variable used in the GSF veto and the $\Delta\eta$ and $\Delta\phi$ variables used for the PF charged hadron veto, for simulation photons passing all the other selection criteria including the pixel veto. To illustrate the effect of each selection, Fig. 4.6 shows the change in the photon selection efficiency and the electron-to-photon fake probability after applying successive

Table 4.3 Effective area values used for the correction of isolation p_T sums of photon objects

$|\eta_{\text{SC}}|$ range	[0, 1]	[1, 1.4442]
$A_{\text{eff}}(I_{\text{CH}})$	0.012	0.010
$A_{\text{eff}}(I_{\text{NH}})$	0.03	0.057
$A_{\text{eff}}(I_{\text{Ph}})$	0.148	0.13

The area is derived in two bins of supercluster η_{SC} for each isolation type. The header of the second and third columns shows the η_{SC} range

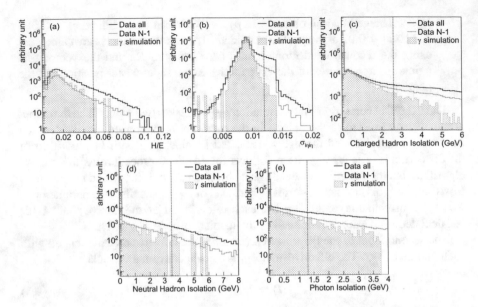

Fig. 4.3 (**a**) H/E, (**b**) $\sigma_{i\eta i\eta}$, (**c**) I_{CH}, (**d**) I_{NH}, and (**e**) I_{Ph} distributions of barrel photon objects with $E_{\text{T}} > 40\,\text{GeV}$ passing the diphoton HLT leading leg requirement. Hollow black histograms are from all objects, while the *blue dotted ones* are the so-called $N-1$ plots. Filled histograms in *light blue* are the distributions taken from a simulation sample. The *red lines* represent the selection cut values

selection requirements, again studied in simulation. The fractions are normalized to the efficiency and the fake probability of the standard electron veto used in CMS for 2012 data (conversion-safe electron veto [3]).

4.4 Electron Selection

Electrons with $p_{\text{T}} > 25\,\text{GeV}$ in the pseudorapidity range $|\eta_{\text{SC}}| < 2.5$ are considered in the analysis, except for those whose superclusters fall in the barrel-endcap gap region $1.4442 < |\eta_{\text{SC}}| < 1.56$.

Similarly to the photons, the electron object is first required to match the trailing-leg HLT object of the diphoton trigger. The match is defined by $\Delta R < 0.15$ between the supercluster position and the trigger object momentum. The ΔR distribution is shown in Fig. 4.2.

The following selection criteria are applied on top of the trigger-matching requirement:

- $H/E < 0.12$
- $\sigma_{i\eta i\eta} < 0.01$ (EB), 0.03 (EE)
- $|d_0| < 0.02\,\text{cm}$, $|d_z| < 0.1\,\text{cm}$

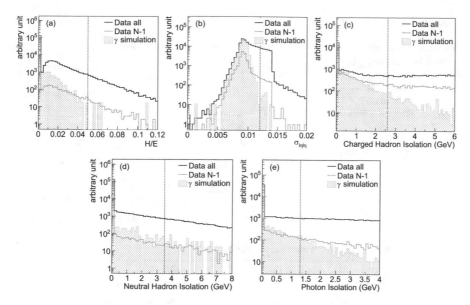

Fig. 4.4 (**a**) H/E, (**b**) $\sigma_{i\eta i\eta}$, (**c**) I_{CH}, (**d**) I_{NH}, and (**e**) I_{Ph} distributions of barrel photon objects with $E_T > 40\,\text{GeV}$ matching the muon plus photon HLT object. See Fig. 4.3 for the meanings of the different histograms

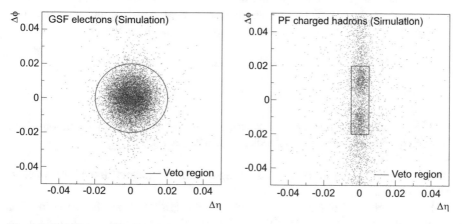

Fig. 4.5 Variables used for the electron vetoes

- $|\Delta\eta_{in}| < 0.06$ (EB), 0.03 (EE):
  The position of the electron track extrapolated from the innermost track state to the ECAL wall is compared to the position of the supercluster. $\Delta\eta_{in}$ represents their difference in η.
- $|\Delta\phi_{in}| < 0.004$ (EB), 0.007 (EE):
  $\Delta\phi_{in}$ is the analogous track-cluster position difference in ϕ.
- $|1/E_{SC} - 1/P_{trk}| < 0.05\,\text{GeV}^{-1}$:

Fig. 4.6 Photon selection efficiency versus electron-to-photon fake probability calculated in photon and electron simulation samples, respectively, plotted for different electron vetoes

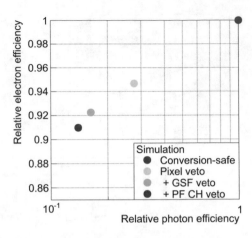

Table 4.4 Effective area values used for the electron isolation correction

$|\eta_{SC}|$ range	[0, 1]	[1, 1.4442]	[1.56, 2]	[2, 2.2]	[2.2, 2.3]	[2.3, 2.4]	[2.4, 2.5]
$A_{eff}(I_{rel})$	0.13	0.14	0.07	0.09	0.11	0.11	0.14

The area is derived in bins of supercluster η_{SC}, whose boundaries are given in the first row

The discrepancy in the energy measured in the ECAL (E_{SC}) and the momentum measured in the tracker (P_{trk}) is an important variable to distinguish an electron-induced signal from an accidental overlap of a track and a cluster.

- $I_{rel} < 0.15$:
  The combined relative isolation I_{rel} is defined as

$$[I_{CH} + \max(0, I_{NH} + I_{Ph} - \rho A^{eff})]/p_T.$$

The isolation p_T sums are all calculated similarly to those for photons. The corresponding effective area A_{eff} is listed in Table 4.4.

- $N_{miss} \leq 1$:
  N_{miss} is the number of missing hits along the track trajectory in the inner tracker.
- Conversion veto
  For each electron object, a probability that it originates from the reconstruction of a photon conversion electron is calculated. This calculation is based on the existence of a nearby compatible track. An electron candidate object must be considered as highly unlikely to be from a conversion.

Figure 4.7 shows the distributions of the variables used in the electron selection, for all trigger-matching electron objects and for the ones that pass all the cuts except the one being plotted. Similar to Figs. 4.3 and 4.4, I_{rel} in the plot is already corrected for PU, and all electrons from period A are used. An electron simulation sample is used for the distributions of the truth-matched electrons for reference. The normalization of the simulation sample is arbitrary.

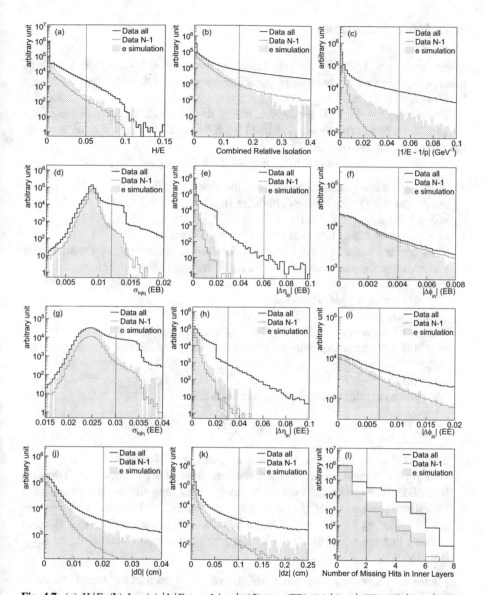

Fig. 4.7 (a) H/E, (b) I_{rel}, (c) $|1/E_{SC} - 1/p_{trk}|$, (d) $\sigma_{i\eta i\eta}$ (EB), (e) $|\Delta\eta_{in}|$ (EB), (f) $|\Delta\phi_{in}|$ (EB), (g) $\sigma_{i\eta i\eta}$ (EE), (h) $|\Delta\eta_{in}|$ (EE), (i) $|\Delta\phi_{in}|$ (EE), (j) $|d_0|$, (k) $|d_z|$, (l) N_{miss} distributions of electron objects with $p_T > 25$ GeV matching the diphoton HLT trailing leg object. See Fig. 4.3 for the meanings of the different histograms

4.5   Muon Selection

Muons with $p_T > 25\,\text{GeV}$ in the pseudorapidity range $|\eta| < 2.4$ are used. The CMS muon system has a full coverage of pseudorapidity up to 2.4, with slight efficiency losses around 0.25 and 0.8.

The muon object is first required to match the muon HLT object of the muon plus photon trigger. The match is defined by $\Delta R < 0.15$ between the momenta of the offline and online objects.

The offline selection of muons comprises the following requirements:

- Muon object is a global muon which is identified in PF:
  For muons with $p_T > 200\,\text{GeV}$, PF identification is not required.
- $|d_0| < 0.2\,\text{cm}$, $|d_z| < 0.5\,\text{cm}$
- Minimum number of hits in inner and outer trackers:
  The muon track must have at least 2 matched muon stations ($N_{\mu\text{-stn.}}$), with at least 1 valid hit in the muon chambers ($N_{\mu\text{-val.}}$). The inner track must have at least 6 tracker layers with hits ($N_{\text{trk.-val.}}$), and at least 1 valid pixel hit (N_{pix}).
- High quality of the global track fit:
  If the muon p_T is less than $200\,\text{GeV}$, χ^2 value of the track fit normalized to the number of degrees of freedom, $\chi^2/N_{\text{d.o.f.}}$, must be less than 10. If p_T is greater than $200\,\text{GeV}$, the uncertainty on p_T must be less than 30% of the measured p_T value.
- $I_{\text{rel}} < 0.12$:
  The combined relative isolation for muons is defined as

$$[I_{\text{CH}} + \max(0, I_{\text{NH}} + I_{\text{Ph}} - 0.5I_{\text{PU}})]/p_T.$$

The isolation p_T sums are calculated in a $\Delta R < 0.4$ cone around the muon. Unlike photons and electrons, no special treatment for the photon isolation I_{Ph} is applied. The charged hadron isolation I_{CH} uses only the particles associated with the muon vertex and is relatively insensitive to PU. The other two p_T sums are corrected by subtracting $0.5I_{\text{PU}}$, where I_{PU} is the sum of the p_T of charged hadrons that are not associated with the primary vertex, calculated in the same isolation cone.

Figure 4.8 shows the distributions of the variables used in the muon selection, for all trigger-matching muons and for the ones that pass all the cuts except the one being plotted. In the plots, $\chi^2/N_{\text{d.o.f.}}$ are plotted only for $p_T < 200\,\text{GeV}$, and $\delta p_T/p_T$ are plotted only for $p_T > 200\,\text{GeV}$. The isolation variable is corrected for PU. The references are taken from muon MC simulation and normalized arbitrarily.

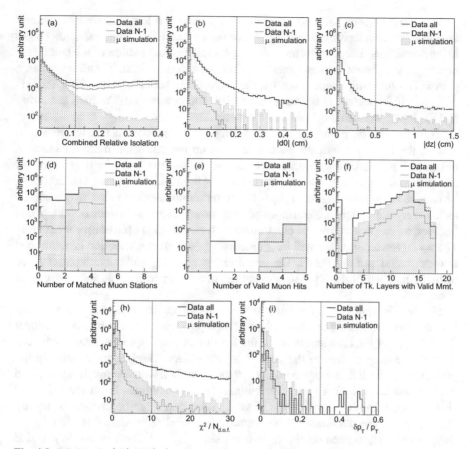

Fig. 4.8 (a) I_{rel}, (b) $|d_0|$, (c) $|d_z|$, (d) $N_{\mu\text{-stn.}}$, (e) $N_{\mu\text{-val.}}$, (f) $N_{trk.\text{-val.}}$, (g) N_{pix}, (h) $\chi^2/N_{d.o.f.}$, and (i) $\delta p_T/p_T$ distributions of muon objects with $p_T > 25$ GeV matching the muon plus photon HLT object. See Fig. 4.3 for the meanings of the different histograms

4.6 Full Selection Criteria

Events with fully selected photons and leptons are further subjected to the requirement that the candidate photon is not near a lepton. This requirement is motivated by two reasons and is implemented in the form of three distinct selection criteria. The first reason is that the Wγ and Zγ MC samples, described in Sect. 4.8, do not simulate events where the photon and the leptons are closer than ΔR of 0.3 and 0.5, respectively. The first selection is therefore to require the candidate photon and lepton with highest p_T to be separated by $\Delta R > 0.8$, so that the search region lies within the simulated phase space. This selection would leave events of Zγ type where the trailing lepton is collinear to the photon. Such events are addressed by the other two requirements described below.

The second reason for the veto is that events where a lepton is near the photon tend to arise from lepton bremsstrahlung or the so-called final-state radiation (FSR) in hard-scattering interactions involving leptons. The cross section of the latter rises rapidly as the photon–lepton system reaches the collinear limit. Therefore, when a photon is found with a lepton nearby, the more likely scenario is that the photon is radiated off the lepton, rather than that the photon comes from the hard scattering and the lepton accidentally points to the same direction.

This observation prompted the remaining two selection criteria, which are to require the leading candidate photon to have no reconstructed muon or electron object with $p_T > 2\,\text{GeV}$ within a cone of $\Delta R < 0.3$, and to require for the $e\gamma$ channel that the invariant mass formed by the leading photon and the leading electron must be outside the window $m_Z \pm 10\,\text{GeV}$. The latter requirement can also be considered as another tool to reject electron-to-photon fakes. There is in fact no sharp line between a fake photon from an electron and a hard radiation off a collinear electron. The no-lepton requirement will be implied whenever photon selection is mentioned in the remainder of this thesis. This requirement also aligns the search phase space to that simulated in the $Z\gamma$ MC simulation sample, as mentioned in the previous paragraph.

Figure 4.9 shows how the latter two vetoes affect the data, the $W\gamma$ and $Z\,\gamma$ background, and the signal simulations. The top panels show the two-dimensional distributions of the invariant mass of the leading photon–lepton system versus ΔR from the leading photon to the nearest lepton object. The bottom panels are the projection of the distributions onto the ΔR axis, overlaid with arbitrarily normalized background and SUSY signal distributions from simulation. From the left-hand side of Fig. 4.9 it is clear that the Z veto removes a large proportion of the $e\gamma$ events, which are dominantly background. The third requirement is effective in the $\mu\gamma$ channel, as seen on the right-hand side. A clear cutoff of the $W\gamma$ and Z γ distributions in the lower panel of both figures illustrates the invalidity of the MC simulation in the low-ΔR region.

Finally, the collected events are compared to the background estimation in the signal region, which is defined by $E_T^{\text{miss}} > 120\,\text{GeV}$ and $M_T > 100\,\text{GeV}$. The transverse mass M_T is defined for each event as $M_T :=$ $\sqrt{2E_T^{\text{miss}}p_T^{\ell}[1 - \cos\Delta\phi(\ell, E_T^{\text{miss}})]}$, where p_T^{ℓ} is the magnitude of the transverse momentum of the leading lepton and $\Delta\phi(\ell, E_T^{\text{miss}})$ is the azimuthal opening angle between the lepton transverse momentum and the E_T^{miss} vector.

Table 4.5 summarizes the event selection requirements along with the number of events surviving each selection.

4.7 Jet Selection

Jets are used later in this analysis to calculate the H_T variable, defined as the sum of the magnitudes of the p_T of the jets. To limit the influence of soft and PU jets, a jet must pass the following criteria to be considered in the H_T sum:

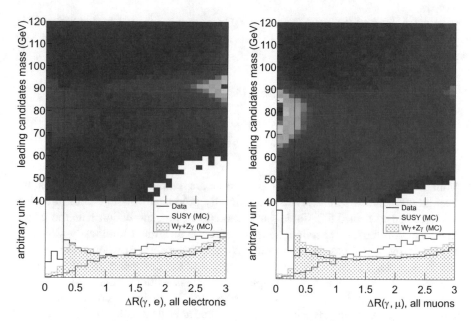

Fig. 4.9 Invariant mass of the leading photon–electron (muon) system versus ΔR from the leading photon to the nearest electron (muon) object, in the $e\gamma$ ($\mu\gamma$) channel data. *Bottom panels* show the projections of the distributions to the ΔR dimension, overlaid with corresponding distributions from Wγ and Z γ background and a SUSY signal simulation. The distributions in the $e\gamma$ ($\mu\gamma$) channel are shown on the *left-(right-)hand side*. For the $e\gamma$ channel, the invariant mass band in $m_Z \pm 10$ GeV is removed from the projections

Table 4.5 Event selection and number of events after successive event selections

Requirement	$e\gamma$ Channel	$\mu\gamma$ Channel
HLT and MET filters	26,733,051	19,456,571
≥ 1 good γ	2,718,364	243,664
≥ 1 good ℓ	70,736	32,173
$\Delta R(\gamma, \ell) > 0.8$	68,168	30,232
Z veto	29,169	–
$E_T^{\text{miss}} > 120$ GeV, $M_T > 100$ GeV	110	152

- $p_T > 30$ GeV
- $|\eta| < 2.5$
- No closer than $\Delta R = 0.5$ from the nearest candidate photon or lepton
- Not tagged as a PU jet by an identification algorithm.

The PU jet identification algorithm [4] in the last criterion calculates the likelihood that a jet originates from a scattering at a non-primary vertex. The likelihood is calculated by a boosted decision tree (BDT) which analyzes variables related to the association of constituent charged particles to the primary vertex and the cluster shape of the jet. A cut on the BDT output is selected so that 87% of PU jets are rejected while 99% of non-PU jets are retained, for a jet p_T between 30 and 50 GeV.

4.8 Simulation

4.8.1 Simulation Data Sets

Multiple MC simulations of SM processes are employed in the analysis for background studies. All data sets are officially produced full-simulation samples, as described in Sect. 3.5.2. Table 4.6 at the end of this subsection lists the name, the generator, the cross section, and the effective integrated luminosity (N/σ) of each data set. The cross section is taken from literature [5, 6] when available. For the $t\bar{t}\gamma$ sample, a k-factor of 2.0 [7] is applied to the cross section calculated by the generator. For WW and WZ samples, NLO cross sections calculated by MCFM are used. For $t\bar{t}$ samples, NNLO cross section of the inclusive production [5] is multiplied by the appropriate branching fraction for each sample. For all other data sets, the value calculated by the generator upon event generation is assigned. To ensure that the high-p_T region of the phase space is sufficiently covered, some data sets are generated separately in bins of p_T of the particle of interest. Such data sets are always used in combination according to their weights (see Eq. (4.1)) in the analysis. For the samples generated with MADGRAPH 5, software versions in the 1.3 family are used, and up to two extra partons are added to the primary process.

The MSSM signal is modeled by three simplified model simulations. In a simplified model, the parameter space of some beyond-the-standard-model (BSM) physics is represented by masses of one or two lightest BSM particles. The other BSM particles that might appear in a realistic scenario are considered decoupled, i.e., too heavy to be directly produced or significantly alter the behavior of the light particles radiatively. Of the three signal models, one features MSSM-like branching ratios, while the other two are further simplified and force the decays of BSM particles exclusively to final states involving a photon and a W boson. The models are called GMSB, TChiwg, and T5wg, with the latter two being exclusive.

The physics description of the signal models are given below:

- GMSB: This is a simplified model of the MSSM where the gluino ($\widetilde{g}$) and the winos ($\widetilde{W}$) are the only sparticles accessible at the LHC. The model phase space is scanned by the MSSM parameters M_2 and M_3, while all other soft masses and the μ parameter are set to 5 TeV. In such a decoupling limit, M_2 and M_3 correspond to physical wino and gluino masses (m_{wino} and m_{gluino}), and the lightest electroweakinos $\widetilde{\chi}^{\pm}_1$ and $\widetilde{\chi}^0_1$ become almost pure winos. An event in the data set can be either a $\widetilde{g}\widetilde{g}$ or $\widetilde{\chi}^{\pm}_1\widetilde{\chi}^0_1$ pair production, illustrated in Fig. 4.10. Processes $\widetilde{\chi}^{\pm}_1\widetilde{\chi}^{\mp}_1$ and $\widetilde{\chi}^0_1\widetilde{\chi}^0_1$ also exist but are not considered since the former does not feature a photon plus lepton final and the latter has a negligible cross

Table 4.6 List of MC simulation data sets for SM processes used in this analysis

Name	Condition	Generator	σ (pb)	$\mathcal{L}_{\text{eff}}$ (fb^{-1})	CMS data set name
Wγ	30 GeV < E_T^γ < 50 GeV	MADGRAPH 5	21.88	39.7	WGToLNuG_PtG-30-50_[8MG]
	50 GeV < E_T^γ < 130 GeV		3.647	311.2	WGToLNuG_PtG-50-130_[8MG]
	E_T^γ > 130 GeV		0.2571	1834.5	WGToLNuG_PtG-130_[8MG]
Zγ	E_T^γ < 130 GeV	MADGRAPH 5	132.6	49.7	ZGToLLG_[8MG]
	E_T^γ > 130 GeV		0.0478	10400	ZGToLLG_PtG-130_[8MG]-pythia6_tauola
$t\bar{t}\gamma$		MADGRAPH 5	2.888	596	TTGJets_[8MG]
WWγ		MADGRAPH 5	0.528	576.3	WWGJets_[8MG]_v2
$t\bar{t}$	Semi-leptonic	MADGRAPH 5	108.7	229.6	TTJets_SemiLeptMGDecays_[8MG]-tauola
	Full-leptonic		26.8	465.6	TTJets_FullLeptMGDecays_[8MG]-tauola
WW		PYTHIA 6.4	56	178.6	WW_[Z2*]_8TeV_pythia6_tauola
WZ		PYTHIA 6.4	33.21	301.1	WZ_[Z2*]_8TeV_pythia6_tauola
W+jets	p_T^W < 50 GeV	MADGRAPH 5	36703.2	1.57	WJetsToLNu_[Z2*]_[8MG]-tarball
	50 GeV < p_T^W < 70 GeV		811.2	59.70	WJetsToLNu_PtW-50To70_[Z2*]_[8MG]
	70 GeV < p_T^W < 100 GeV		428.9	52.34	WJetsToLNu_PtW-70To100_[Z2*]_[8MG]
	p_T^W > 100 GeV		228.9	55.67	WJetsToLNu_Ptw-100_[Z2*]_[8MG]
Drell-Yan		MADGRAPH 5	2950.0	10.33	DYJetsToLL_M-50_[Z2*]_[8MG]-tarball
Drell-Yan	Fastsim	MADGRAPH 5	2950.0	10.33	DYJetsToLL_M-50_[Z2*]_[8MG]-tarball
QCD	30 GeV < $\hat{p}_T$ < 50 GeV	PYTHIA 6.4	6.63×10^7	9.05×10^{-2}	QCD_Pt-30to50_[Z2*]_8TeV_pythia6
	50 GeV < $\hat{p}_T$ < 80 GeV		8.15×10^6	7.36×10^{-1}	QCD_Pt-50to80_[Z2*]_8TeV_pythia6
	80 GeV < $\hat{p}_T$ < 120 GeV		1.03×10^6	5.83	QCD_Pt-80to120_[Z2*]_8TeV_pythia6
	120 GeV < $\hat{p}_T$ < 170 GeV		1.56×10^5	38.5	QCD_Pt-120to170_[Z2*]_8TeV_pythia6
	170 GeV < $\hat{p}_T$ < 300 GeV		3.42×10^4	175	QCD_Pt-170to300_[Z2*]_8TeV_pythia6

In the table, E_T^γ and p_T^W are the p_T of the photon and W boson, respectively, and $\hat{p}_T$ is the hardest parton. Semi-leptonic and full-leptonic decays refer to the cases where one or two W bosons from the top quark decay leptonically. The abbreviations in the data set names are as follows: [8MG] = 8TeV-madgraph; [Z2*] = Tunez2star or Tunez2Star

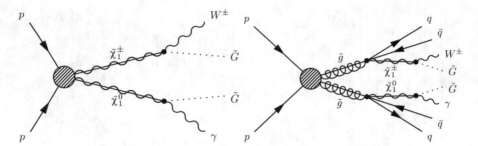

Fig. 4.10 $\widetilde{\chi}^{\pm}{}_1\widetilde{\chi}^0{}_1$ (*left*) and $\widetilde{g}\widetilde{g}$ (*right*) productions in the GMSB model which leads to the photon plus lepton final-state signature

section. The cross section of processes mediated by the sleptons, squarks, and other sparticles is also negligible. The only kinematically allowed decays of $\widetilde{g}$ are either to $q\bar{q}\widetilde{\chi}^{\pm}{}_1$ or $q\bar{q}\widetilde{\chi}^0{}_1$. Similarly, the $\widetilde{\chi}^{\pm}{}_1$ decay is limited to $W^{\pm}\widetilde{G}$ and $\widetilde{\chi}^0{}_1$ to $\gamma\widetilde{G}$ or $Z\widetilde{G}$. The mass parameter space of $715\,\text{GeV} \leq m_{\text{gluino}} \leq 1615\,\text{GeV}$ and $205\,\text{GeV} \leq m_{\text{wino}} \leq (m_{\text{gluino}} - 10\,\text{GeV})$ is sampled with $50\,\text{GeV}$ steps in both m_{wino} and m_{gluino}.

- TChiWg: This is a model initiated by the direct production of gaugino-like particles $\widetilde{\chi}^{\pm}$ and $\widetilde{\chi}^0$. The decays of $\widetilde{\chi}^{\pm}$ and $\widetilde{\chi}^0$ are restricted to $W^{\pm}\widetilde{G}$ and $\gamma\widetilde{G}$, respectively, where $\widetilde{G}$ is a nearly massless BSM particle. Thus the model closely resembles the electroweakino production in the GMSB model, but with a fixed neutralino decay. A mass range of $100\,\text{GeV} \leq m_{\widetilde{\chi}} \leq 800\,\text{GeV}$, where $m_{\widetilde{\chi}}$ is the degenerate mass of $\widetilde{\chi}^{\pm}$ and $\widetilde{\chi}^0$, is sampled in $10\,\text{GeV}$ steps.

- T5Wg: This is a model designed after the strong-production mode of the GMSB model. In each event, a pair of gluino-like colored particles is produced. The symbol $\widetilde{g}$ is also used to describe this particle when discussing this simplified model. Each $\widetilde{g}$ decays either to $q\bar{q}\widetilde{\chi}^{\pm}$ or $q\bar{q}\widetilde{\chi}^0$ with equal likelihood, but the $\widetilde{g}\widetilde{g}$ decays to $\widetilde{\chi}^{\pm}\widetilde{\chi}^{\mp}$ and $\widetilde{\chi}^0\widetilde{\chi}^0$ are not considered because they do not lead to a photon plus lepton final state. The samples are generated varying the mass of $\widetilde{g}$ ($m_{\widetilde{g}}$) and $m_{\widetilde{\chi}}$. The phase space of $400\,\text{GeV} \leq m_{\widetilde{g}} \leq 1350\,\text{GeV}$ and $25\,\text{GeV} \leq m_{\widetilde{\chi}} \leq (m_{\widetilde{g}} - 25\,\text{GeV})$ is scanned with $50\,\text{GeV}$ steps in the two mass parameters.

The GMSB model data set is privately produced and generated exclusively with PYTHIA 6.4. The input mass spectrum for each mass point is written in a SUSY Les Houches Accord (SLHA) [8] file produced by SuSpect2.41 [9] and SDECAY1.3 [10]. The former program calculates the mixing matrices given the input mass spectrum at the weak scale. Its output is passed to the latter, where the branching ratios are derived from the mixings. The mass of $\widetilde{G}$ is not specified in the SLHA. The "GMSB" option in PYTHIA is turned on to simulate the NLSP decays to $\widetilde{G}$ and its SM partner. For each mass point, 25,000 $\widetilde{g}\widetilde{g}$ production and 12,500 $\widetilde{\chi}^{\pm}{}_1\widetilde{\chi}^0{}_1$ production events are generated. The two production modes are processed independently because their cross sections are calculated separately, and later combined into a single data set.

The generation of the TChiWg and T5Wg data sets, which are semi-official, employs both MADGRAPH 5 and PYTHIA 6.4. The initial pair production of the BSM particles is simulated with MADGRAPH 5 version 1.5.4 using the mssm model, with up to two additional partons. To run with the mssm model, $\widetilde{\chi}^{\pm}$ and $\widetilde{\chi}^0$ were identified with the winos and $\widetilde{g}$ with the gluino. The subsequent decays of $\widetilde{\chi}^{\pm}$, $\widetilde{\chi}^0$, and $\widetilde{g}$ were simulated by PYTHIA. The input to MADGRAPH is a mass spectrum also written in a SLHA file for each parameter point, but mixings and branching ratios are not calculated since the generation did not involve any decay. The output Les Houches Events (LHE) [11] files from MADGRAPH 5, which list the four-momenta of the pair-produced sparticles and additional partons, are then processed by PYTHIA 6.4 to simulate the isotropic decay of the BSM particles. The events with the decayed BSM particles are once again written out to LHE files. These LHE files are passed to the CMS central MC production machinery to generate all remaining decays and the parton shower with PYTHIA, and then run the detector simulation. Approximately 60,000 events are generated for each mass point of the TChiWg and T5Wg data sets.

The cross sections for the GMSB points are calculated with PROSPINO2 [12] by running over the input SLHA files. The factorization scale μ_F was set to m_Z in the calculation. The calculation accuracy of PROSPINO2 is NLO in QCD. For TChiWg and T5Wg, the cross sections calculated by the LHC SUSY cross section working group [13] for chargino–neutralino production and gluino pair-production simplified models are assigned. The gluino pair-production cross sections are scaled by a factor 0.5, accounting for the ignored $\widetilde{g}\widetilde{g}$ decays to $\widetilde{\chi}^{\pm}\widetilde{\chi}^{\mp}$ and $\widetilde{\chi}^0\widetilde{\chi}^0$. Figure 4.11 shows the cross sections assigned to the three models.

For the GMSB model, the branching fraction to the photon plus lepton final state is dependent on the gluino and wino masses. Since the branching fraction differs also by the SUSY production mechanism, for each mass point, an effective branching fraction b is calculated as

$$b = \frac{\left(\sum_P \sigma_P N_P^{\gamma\ell}\right)^2}{\sum_P \sigma_P^2 N_P^{\gamma\ell}} \Big/ \frac{\left(\sum_P \sigma_P N_P^{\text{gen.}}\right)^2}{\sum_P \sigma_P^2 N_P^{\text{gen.}}}, \tag{4.4}$$

where $P \in \{\widetilde{g}\widetilde{g}, \widetilde{\chi}^{\pm}_1 \widetilde{\chi}^0_1\}$ is the production process, σ_P is the cross section of the process for the mass point, $N_P^{\gamma\ell}$ is the number of events with a photon plus lepton final state, and $N_P^{\text{gen.}}$ is the number of generated events. The effective branching fractions for the $e\gamma$ channel are also shown in Fig. 4.11. The values for the $\mu\gamma$ channel are identical. The corresponding fractions for the TChiWg and T5Wg models are trivial and are given by the leptonic branching fractions of the W boson.

Unlike the SM background data sets, the detector response to the signal models are simulated by the CMS fast simulation described in Sect. 3.5.2. The simulated samples are then subjected to the full event selection criteria applied to data to estimate how many signal events are expected in the signal region (see Sect. 4.6) for each mass point.

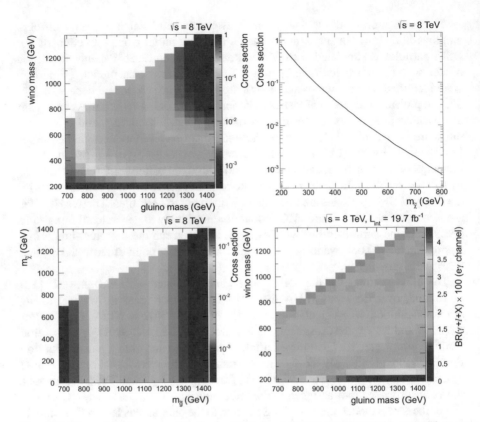

Fig. 4.11 The assigned cross sections for the GMSB (*top left*), TChiWg (*top right*), T5Wg (*bottom left*) model points. The branching fractions to the photon plus lepton final state in the eγ channel of the GMSB model are shown at the *bottom right*

4.8.2 Pileup Reweighting

All of the MC simulation samples described above include multiple simulations of minimum bias scattering in each event together with the primary physics process. The additional interactions model the PU. For full-simulation samples, PU from the neighboring bunch crossings as well as the event bunch crossing are injected. To simulate a realistic effect of PU, the number of additional interactions is randomly chosen from a predetermined distribution, called the PU profile. Because the PU profile had to be predicted for the production campaign of the MC simulation data sets before the 2012 run started, there is a discrepancy between the observed and simulated PU profiles. The PU profile used in the data sets listed in Sect. 4.8.1 is called S10.

Since PU plays an important role in many aspects of the object reconstruction and selection, a correction is applied to the simulation samples by assigning each event a weight determined by the number of interaction vertices that it is generated

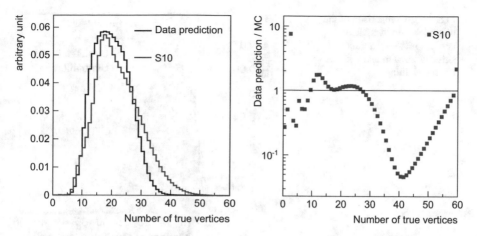

Fig. 4.12 *Left*: Predicted number of vertices in data and true number of generated vertices in simulation with S10 profile. *Right*: Data-to-simulation ratio

with. While it is simple to compare the number of reconstructed vertices between data and MC simulation, doing so will lead to a bias, since the vertex reconstruction efficiency itself depends on the PU activity in the event. Therefore, to compare the profiles that are as unbiased as possible, the number of generated vertices in the MC simulation ($N_{\mathrm{vtx}}^{\mathrm{MC,true}}$) and the predicted number of inelastic scatterings in data ($N_{\mathrm{vtx}}^{\mathrm{data,pred.}}$) are used to derive the weights. The prediction in data is made for every 23 s interval of the 2012 run using the instantaneous luminosity information and the total pp inelastic scattering cross section of the proton of 69.4 mb, which is an extrapolated value from the $\sqrt{s} = 7\,\mathrm{TeV}$ measurements [14, 15].

The distributions of $N_{\mathrm{vtx}}^{\mathrm{data,pred.}}$ and $N_{\mathrm{vtx}}^{\mathrm{MC,true}}$, as well as their ratio, are shown in Fig. 4.12. Each MC simulation event is weighted by the ratio value corresponding to the $N_{\mathrm{vtx}}^{\mathrm{MC,true}}$ of the event. As a cross-check of the reweighting method, distributions of the number of reconstructed vertices in $Z \to \mu\mu$ events for data and MC simulation are compared, with and without reweighting applied to the simulation. Figure 4.13 displays this comparison, which shows that the reweighting clearly brings the simulation distribution closer to that in data.

4.8.3 Efficiency Corrections

To compare the MC simulation with data, differences in trigger and object selection efficiencies must be accounted for. In a similar manner to the PU reweighting, the MC simulation events are given weights according to the data-to-simulation efficiency scale factors (ESF). For each simulation event, the weight w is calculated as

$$w = \prod_{j \in \mathrm{pass}} R_j \cdot \prod_{k \in \mathrm{fail}} \frac{1 - R_k \epsilon_k^{\mathrm{MC}}}{1 - \epsilon_k^{\mathrm{MC}}}, \tag{4.5}$$

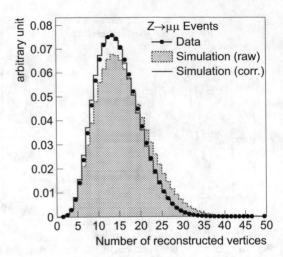

Fig. 4.13 Comparison of the number of reconstructed vertices in $Z \rightarrow \mu\mu$ data and simulation, with and without PU reweighting

where ϵ^{MC} is the full selection efficiency, i.e., the probability of firing the trigger and passing the offline selection, of simulated photons and leptons. R_j is the ESF, defined as $R_j = \epsilon_j^{data}/\epsilon_j^{MC}$ with data efficiency ϵ^{data}. Indices j and k run over all photons and leptons in the event, where j is used for the objects that pass the selection criteria and k for those that fail. The first factor in Eq. (4.5) represents the efficiency corrections, while the second factor gives the inefficiency corrections.

In general, the selection and trigger efficiencies depend on the event properties such as the amount of hadronic activity, and thus differ slightly from sample to sample. The inherent assumption in this formula is that the variation of the efficiency values under such effects is proportional between data and simulation, so that the ESF stays constant if events with similar properties are compared.

The calculation of efficiency values requires a sample of pure objects. In MC simulation, such a sample is trivially prepared because it is possible to know from the MC truth information which particle caused the detector hits of the reconstructed object. Thus the derivation of the ESF amounts to a measurement of ϵ^{data} using various techniques. The efficiency measurements are discussed in Sect. 5.6.

The ESF values for the signal simulation samples are different from those in the background samples, because the former is produced with the CMS fast simulation, which has an inferior modeling of the detector response. Therefore, correction factors are applied to the ESF values for the signal samples. These factors are obtained by comparing the Drell-Yan (DY) and DY Fastsim data sets, which are simulations of identical physics processes with the full and fast simulation.

References

1. CMS Collaboration: CMS Luminosity Based on Pixel Cluster Counting - Summer 2013 Update. CMS Physics Analysis Summary CMS-PAS-LUM-13-001. CERN (2013). http://cdsweb.cern.ch/record/1598864
2. Cacciari, M., Salam, G.P., Soyez, G.: The catchment area of jets. J. High Energy Phys. **0804**, 005 (2008). LPTHE-07-02. doi:10.1088/1126-6708/2008/04/005. arXiv: 0802.1188 [hep-ph]
3. CMS Collaboration: Search for a Higgs boson decaying into two photons in the CMS detector. Tech. rep. CMS-PAS-HIG-11-010 (2011)
4. CMS Collaboration: Pileup Jet Identification. CMS Physics Analysis Summary CMS-PAS-JME-13-005 (2013). http://cdsweb.cern.ch/record/1581583
5. Czakon, M., Fiedler, P., Mitov, A.: Total top-quark pair-production cross section at hadron colliders through $O(\alpha_S^4)$. Phys. Rev. Lett. **110**, 252004 (2013). doi:10.1103/PhysRevLett.110.252004. arXiv: 1303.6254 [hep-ph]
6. Gavin, R., Li, Y., Petriello, F., Quackenbush, S.: FEWZ 2.0: a code for hadronic Z production at next-to-next-to-leading order. Comput. Phys. Commun. **182**, 2388–2403 (2011). ANL-HEP-PR-10-60. doi:10.1016/j.cpc.2011.06.008. arXiv: 1011.3540 [hep-ph]
7. Melnikov, K., Schulze, M., Scharf, A.: QCD corrections to top quark pair production in association with a photon at hadron colliders. Phys. Rev. D **83**, 074013 (2011). doi:10.1103/PhysRevD.83.074013. arXiv: 1102.1967 [hep-ph]
8. Allanach, B., et al.: SUSY Les Houches Accord 2. Comput. Phys. Commun. **180**, 8–25 (2009). FERMILAB-PUB-07-036-T, SLAC-PUB-12765, CERN-PH-TH-2007-148, DAMTP-2007-76, EDINBURGH-2007-31, KEK-TH-1170, LAPTH-1204-07, LPT-ORSAY-07-81, SHEP-07-13. doi:10.1016/j.cpc.2008.08.004. arXiv: 0801.0045 [hep-ph]
9. Djouadi, A., Kneur, J.-L., Moultaka, G.: SuSpect: a fortran code for the supersymmetric and Higgs particle spectrum in the MSSM. Comput. Phys. Commun. **176**, 426–455 (2007). PM-02-39, CERN-TH-2002-325. doi:10.1016/j.cpc.2006.11.009. arXiv: hep-ph/0211331 [hep-ph]
10. Muhlleitner, M., Djouadi, A., Mambrini, Y.: SDECAY: a fortran code for the decays of the supersymmetric particles in the MSSM. Comput. Phys. Commun. **168**, 46–70 (2005). CERN-TH-2003-252, PM-03-22, PSI-PR-03-17. doi:10.1016/j.cpc.2005.01.012. arXiv: hep-ph/0311167 [hep-ph]
11. Alwall, J., et al.: A standard format for Les Houches event files. Comput. Phys. Commun. **176**, 300–304 (2007). FERMILAB-PUB-06-337-T, CERN-LCGAPP-2006-03. DOI: 10.1016/j.cpc.2006.11.010. arXiv: hep-ph/0609017 [hep-ph]
12. Beenakker, W., Hopker, R., Spira, M.: PROSPINO: a program for the production of supersymmetric particles in next-to-leading order QCD. arXiv: hep-ph/9611232 (1996)
13. LHC SUSY Cross Section Working Group. https://twiki.cern.ch/twiki/bin/view/LHCPhysics/SUSYCrossSections. Retrieved 1 Jan 2015
14. CMS Collaboration: Measurement of the inelastic proton-proton cross section at $\sqrt{s} = 7$ TeV. Phys. Lett. B **722**, 5–27 (2013). CMS-FWD-11-001, CERN-PH-EP-2012-293. doi:10.1016/j.physletb.2013.03.024. arXiv: 1210.6718 [hep-ex]
15. TOTEM Collaboration: Measurement of proton-proton inelastic scattering cross-section at $\sqrt{s} = 7$ TeV. Europhys. Lett. **101**, 21003 (2013). doi:10.1209/0295-5075/101/21003

Chapter 5
Data Analysis

5.1 Outline

The principal concern of the data analysis in this SUSY search is to accurately estimate how many of the candidate events selected in the signal region as described in Sect. 4.6 can be attributed to known SM processes. Multiple SM processes, or backgrounds, can contribute events in the signal region. There is no way to distinguish the SUSY signal from the SM backgrounds on an event-by-event basis, since the signal region is by definition where all the cuts have been applied. Therefore, the background estimation must be done statistically.

Three categories of SM backgrounds are considered. The first is the fake photon category, where the photon objects in the candidate sample are not physical prompt photons. This category is further split into two sub-categories: one in which the photon object is faked by an electron; and one where it is faked by hadrons. The second is the fake lepton category, where the lepton candidate objects arise from hadronic activities. The last background category is the electroweak (EWK), which is dominantly SM $W\gamma$ and $Z\gamma$ productions. In particular, leptonically decaying $W\gamma$ events have the final state of an energetic photon, a lepton, and significant E_T^{miss} from the escaping neutrino, which is exactly the signature investigated in this search. The two backgrounds are collectively called the $V\gamma$ background. Subdominant components of the EWK category are the rare multiboson and top quark production, which have minor contributions in total but are not negligible in the high-E_T^{miss} region. The so-called double fake events, where both the photon and the lepton objects are the results of mis-identification, are accounted for consistently among the fake background estimation methods.

The two types of fake photon backgrounds are estimated from control samples in real data, as opposed to simulation. Since the interest is in not only the amount of backgrounds but also their distributions in the kinematic variables such as E_T^{miss}, the following method has been developed. First, a control sample enriched in electrons

© Springer International Publishing AG 2017
Y. Iiyama, *Search for Supersymmetry in pp Collisions at* $\sqrt{s} = 8$ TeV *with a Photon, Lepton, and Missing Transverse Energy*, Springer Theses,
DOI 10.1007/978-3-319-58661-8_5

or jets that fake the photons is prepared. This control sample is called the proxy sample. The selection of events for the proxy sample is such that the proxy objects, i.e., the electrons or jets, have similar kinematic properties to that of the faking objects in the signal candidate events. The events in the proxy sample are then assigned weights so that their weighted distributions in the kinematic variables of interest predict the distributions of the fake photon events in the signal region, both in shape and in normalization. The event weights are called transfer factors, and are measured from real data. Sections 5.2 and 5.3 describe the fake photon background estimations.

The fake lepton background is also estimated from real data using a control sample, but without determining its absolute normalization. Instead, the normalization is given together with that for the $V\gamma$ background, whose distribution shapes are taken from MC simulation. Template fits in a low-E_T^{miss} control region are used to determine the predicted rates for the two backgrounds simultaneously. Lastly, the rare backgrounds are fully simulation-based and are normalized to the theoretical cross sections. The fake lepton and EWK background estimations are described in Sects. 5.4 and 5.5.

In the remainder of this chapter, real data samples are simply called data. Since the analysis involves comparing simulated background samples to data, the data-to-simulation ESF must be determined and applied to the simulation samples. As mentioned in Sect. 4.8.3, the nontrivial parts of the ESF determination are the efficiency measurements on data, which are described in Sect. 5.6.

Throughout this analysis, events with $E_T^{miss} > 70\,\mathrm{GeV}$ are excluded in the background estimation methods whenever the potential signal contribution can bias the result. This cut is determined by observing that one T5wg point, with $m_{\tilde{g}} = 500\,\mathrm{GeV}$ and $m_{\tilde{\chi}} = 425\,\mathrm{GeV}$, has 95% of its events in the $E_T^{miss} > 70\,\mathrm{GeV}$ region. Since this is a relatively low-mass point, it is expected that this fraction is even higher for more relevant higher-mass points.

5.2 Fake Photon Background Due to Electrons

Events with an electron faking a photon are relevant in the $e\gamma$ channel, where the identified electron and photon candidates can in fact be two electrons from DY dielectron production. Additionally, in both channels, events from fully leptonic $t\bar{t}$ decays resulting in $e\mu$ or ee have minor contributions to the tail of the E_T^{miss} distribution, if one of the electrons fake the photon. The data-driven estimation of the background contribution from such sources is described below. The full details of the estimation for the $e\gamma$ channel are presented first. The procedure in the $\mu\gamma$ channel is almost identical and is described at the end of this section.

5.2.1   Electron Proxy Sample Definition (eγ)

The proxy sample for the eγ channel is obtained by replacing the photon object in the signal candidate sample by an electron proxy object, while keeping all the other selection criteria unaltered. The electron proxy object, in turn, is a photon object which is selected by a set of conditions identical to the candidate photon selection, except for the pixel and GSF veto requirements, which are inverted. In other words, the electron proxy must fail either one or both of the pixel and GSF veto, but still satisfy the PF charged hadron veto. The reason for not inverting the latter is as follows. The electron proxy object is supposed to be a reconstruction of a physical electron. Having a PF charged hadron matched to it indicates a mis-reconstruction in the PF algorithm, which adversely affects the resolution of other quantities calculated from PF objects, in particular E_T^{miss}. Thus, events where the photon candidate fails the PF charged hadron veto are also disregarded from the electron proxy sample.

When more than one electron proxy object can be identified in the event with an accompanying electron candidate, all proxy-electron pairs are used in the background estimation independently as if they come from two different events. Such a case typically arises when the two electrons from a Z boson decay are both in the barrel and have sufficiently high p_T. Since the electron proxy definition and the candidate electron selection are not mutually exclusive, it is possible for the e^+e^- pair to register as proxy-electron combinations in two ways. This apparent double-counting is intentional; since the proxy-to-fake transfer factor represents how many electron fakes are expected per electron proxy object, every proxy configuration that would become a candidate event by replacing the electron proxy with a candidate photon must be counted. In other words, conceptually, the number of fake photons is estimated as the number of electron proxy objects, instead of the number of events, that appear in the electron proxy sample, multiplied by the transfer factor.

5.2.2   Transfer Factor Measurement (eγ)

To determine the transfer factor $R = N_f/N_p$, where N_f is the number of fake photons arising from electrons and N_p the number of electron proxy objects, the electron fake rate f, defined as

$$f = \frac{N_f}{(N_p + N_f)},$$

(5.1)

is measured first. The fake rate is trivially related to the transfer factor by $R = f/(1-f)$.

The fake rate, and therefore the transfer factor, is determined as a function of several event variables such as the p_T of the electron. Each event in the proxy sample receives a weight that depends on the values of these variables in the event. The use of a functional transfer factor, as opposed to a scaling of the entire proxy sample by a single factor, is motivated from the observation that different simulation data sets exhibit starkly different overall fake rates. The fake rates ,however, agree when looked in fine bins of certain variables, i.e., seen as a function of these variables. Thus the reason for the disagreement is that the distributions in these variables differ between the data sets, due to the difference in the simulated physics process. For example, $Z \rightarrow ee$ process has a strongly peaking electron p_T distribution, while that of the $t\bar{t}$ process is smooth but has a long tail towards high p_T. When a function of p_T is integrated together with such two different distributions to obtain an average in the sample, the result can differ significantly. The implication is that the overall fake rate, which is such an integral of the fake rate function, is not reliable when it is measured in one sample but applied to another. Since the fake rate is measured on a $Z \rightarrow ee$ sample as described below but is also used to estimate the $t\bar{t}$ background in the signal region, the underlying fake rate function is needed for an accurate background estimation.

The following three variables are chosen as the argument to the fake rate function: the transverse momentum of the fake or proxy electron (p_T^e); the number of tracks associated with the primary vertex (N_{trk}); and the number of reconstructed vertices in the event (N_{vtx}). The first two variables are chosen to factor out the differences between physics processes, as mentioned above. The number of vertices, on the other hand, was used to mitigate the mis-modeling of the background pileup distribution, which in turn can cause mis-estimation of the resolutions of quantities such as E_T^{miss}. The second variable, N_{trk}, is not a commonly studied variable; its significance in characterizing the event is described in Sect. 5.2.7. Figure 5.1 shows the distributions of these three variables for $Z \rightarrow ee$ and $t\bar{t}$ simulations.

The fake rate measurement employs the tag-and-probe (TP) technique on $Z \rightarrow ee$ decay. In a TP measurement, the decay products of a well-identified resonance such as the Z boson is utilized to study the selection efficiency of various object selection criteria, such as the photon candidate selection and the electron proxy

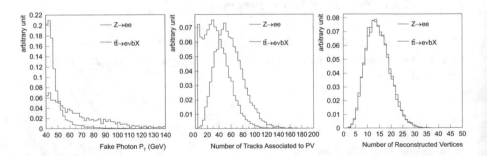

Fig. 5.1 Distributions of the p_T^e (*left*), N_{trk} (*center*), and N_{vtx} (*right*) in two simulation samples representing the most relevant physics processes. Only photon objects matched to a generator-level electron were used for the p_T^e distribution, and only events containing such photons are used for the other two distributions

selection, on one of the decay products, such as an electron from a $Z \to ee$ decay. The reconstructed object (e.g., a photon object) where the different object selections are tried is called the probe. All decay products except for the one corresponding to the probe must be well-identified, and are collectively called the tag. From the invariant mass distribution of the tag-probe system around the resonance mass, the number of resonance decay events can be estimated. The ratio of the selection efficiencies of the different selection criteria can then be inferred from the ratio of the estimates of the numbers of resonance decays with different probe definitions.

For the fake rate measurement, the TP method is performed on a sample (TP sample) of dielectron events with one electron object, the tag, passing the selection criteria in Sect. 4.4. The sample is taken from the Photon/DoublePhoton data set in the control region $E_{\mathrm{T}}^{\mathrm{miss}} < 70\,\mathrm{GeV}$. The event selection in fact follows that of the $e\gamma$ channel closely. Besides the electron, a photon object passing the photon candidate selection excluding the pixel and GSF veto requirements must be present in the event. This photon object is the probe in the measurement. The E_{T} threshold of the probe is lowered to $25\,\mathrm{GeV}$ to allow for a wider E_{T} range for this measurement. This implies that the probe is not always matching the leading leg of the diphoton trigger, which is one of the photon selection criteria in Sect. 4.3, and instead it is only required that either of the tag or the probe matches it. Both objects still have to match the trailing leg. Additionally, for a probe with $E_{\mathrm{T}} < 40\,\mathrm{GeV}$, the tag p_{T} has to be greater than $40\,\mathrm{GeV}$ to ensure that the measurement is performed in a kinematic region that the diphoton trigger is fully efficient. The sample thus prepared constitutes the denominator sample of the fake rate measurement; its subset where the probe passes the full photon selection criteria is the numerator sample.

To derive the fake rate as a function of $p_{\mathrm{T}}^{\mathrm{e}}$, N_{trk}, and N_{vtx}, the full TP sample is binned in these variables and a fake rate measurement is performed in each bin separately. The bins in which the individual measurement took place are referred to as TP bins in the remainder of this section. Assuming that the 2-dimensional $(p_{\mathrm{T}}^{\mathrm{e}}, N_{\mathrm{trk}})$ distribution and the 1-dimensional N_{vtx} distribution do not correlate, the same data set is binned in two different ways, once in the first two variables and then in N_{vtx}. The basis of this assumption is discussed in Sect. 5.2.5. In total, 119 TP bins are analyzed.

5.2.3   Individual Fake Rate Measurement ($e\gamma$)

In each TP bin, the fake rate given in Eq. (5.1) can be calculated by estimating N_{f} and $N_{\mathrm{p}} + N_{\mathrm{f}}$ from the number of $Z \to ee$ events in the numerator and denominator. This estimate is obtained from a fit to the dielectron invariant mass distribution around m_{Z}. The integral of the function that describes the m_{Z} distribution gives the number of Z boson decays present in the sample.

The invariant mass fit is performed using a probability distribution function which comprises two templates, one for the signal (Z boson decay) and another to model the background. The signal template is constructed from simulation events by applying the denominator sample selection to the DY data set and taking the invariant mass distribution. In addition, to be used for the signal template, the tag and probe objects are required to each match by $\Delta R < 0.1$ to a generator-level final-state electron that originates from a Z boson. To form the probability distribution function of the invariant mass distribution, a Gaussian adaptive kernel estimator [1] is employed whenever computationally viable. Otherwise, i.e., when there are abundant events to form the distribution, a histogram-based template with a second-order interpolation is used. One signal template is produced for each TP bin, and is used for each fit on both the denominator and numerator sample. To account for detector resolution effects and loss of electron energy due to radiation, the template is convoluted with a Gaussian with the mean and width floating for each fit.

The background template is taken from the MuEG data set. The underlying idea is that the background in the fit sample is dominantly due to processes that involve a real electron, such as $W\gamma \rightarrow e\nu\gamma$ and $Z \rightarrow ee\gamma$. Such processes are lepton-flavor symmetric, i.e., have virtually identical rate for electronic and muonic events. Thus, the shape of the background can be approximated by the invariant mass distribution formed by a muon and the probe. The details of the background template preparation are as follows. As in the event selection for the $\mu\gamma$ channel, events with at least one muon with $p_T > 25\,\text{GeV}$ matching the muon leg of the trigger are required. The probe definition is identical to the TP sample, except for the trigger requirement. Since it is not possible to require the probe to match the diphoton trigger object in this case, the offline isolation criteria $I_{ECAL} < 5\,\text{GeV}$, $I_{HCAL} < 5\,\text{GeV}$, and $I_{Trk} < 5\,\text{GeV}$ are applied in place of the trigger-matching condition. The background probability distribution function is constructed from this muon-probe sample by the Gaussian kernel estimator mentioned above.

The floating parameters in the fit are the normalization factors of the templates and the signal smearing and shifting parameters. Extended maximum-likelihood fits are performed for the denominator and numerator samples. An example of such fits is shown in Fig. 5.2.

Once the shape and normalization of the signal template are determined, the resulting signal mass distribution is integrated from 60 to 120 GeV. The fake rate of the TP bin is then calculated as the ratio between the integral in the numerator sample to that in the denominator sample. Figure 5.3 shows the calculated fake rate in each TP bin, represented as points in the plots in the panels on the left-hand side. The error bars on the points include the systematic uncertainties, discussed next.

5.2.4 Systematic Uncertainties on Tag-and-Probe (eγ)

Since each fake rate value is essentially an inefficiency fraction, where the statistical fluctuations of the numerator and denominator are in principle correlated, the Clopper-Pearson 1σ interval is assigned as its statistical uncertainty. This treatment is equivalent to assuming that the number of Z decays is counted perfectly, both in

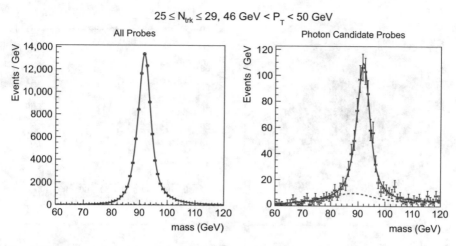

Fig. 5.2 Fits to the Z boson mass peak for the determination of the electron mis-identification rate. The *black points* with uncertainty bars, the *blue solid line*, and the *blue dashed line* represent the target distribution to be fit, the full fit model, and the background model, respectively. *Left*: denominator sample for $46\,\text{GeV} < p_T^e < 50\,\text{GeV}$ and $25 \le N_{trk} \le 29$. *Right*: numerator sample in the same p_T^e and N_{trk} bin

the numerator and the denominator. In reality, however, the fit method has certain inaccuracies, which will manifest themselves as systematic errors on the counts. A full evaluation of the uncertainty would therefore require a convolution of the systematic uncertainty with the statistical one. However, it is observed that the full convolution gives a result similar to what is obtained by simply adding the two in quadrature. For the sake of simplicity, the latter method is used to combine the statistical and systematic uncertainties.

The mis-modeling of both the signal and the background templates are the major systematic errors. The effect of each error is evaluated independently. In addition, the fake rate measurement is tested on MC simulation to account for other residual errors.

To assess the effect of the signal template mis-modeling, the fit procedure is performed with no Gaussian convolution to the signal template. The resulting fake rates differed from the nominal values by 1–2%. The effect of background mis-modeling is estimated using a modified background template, which employed a mixture of DY and W+jets MC simulation samples. In such a mixture, the shape of the background template obtained by the muon-probe selection is compared to the shape of the true background, i.e., tag-probe events where the probe is not matched to a generator-level electron. The top panels of Fig. 5.4 show both distributions. The contributions from the two simulation data sets are indicated separately. The distributions are overlaid in the bottom-left panel of the same figure. The ratio of the two distributions is then multiplied to the data muon-probe mass distribution, thus modifying its shape. The TP fit is repeated with this alternative background template, yielding fake rates that are different from the nominal values by about 4%. The right bottom panel of Fig. 5.4 shows an example of the muon-probe mass distributions in data before and after this modification.

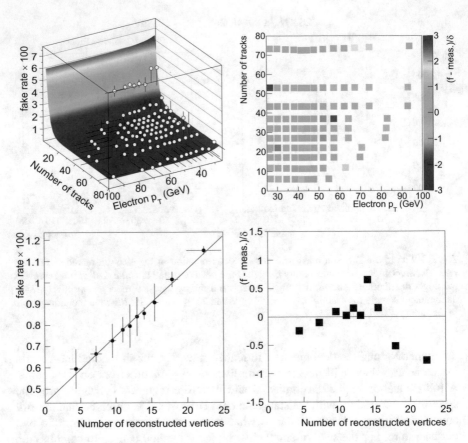

Fig. 5.3 *Left*: Measured fake rate in $(p_T{}^e, N_{trk})$ *(top)* and N_{vtx} *(bottom)* bins. The fake rate function, discussed in Sect. 5.2.5, are overlaid, respectively, as a *colored surface* and a *line* to the measured data points. *Right*: Normalized fit residuals of the fake rate function

Additionally, to account for residual effects, the full fake rate determination procedure is performed on a MC simulation sample mixture mentioned above. The discrepancy between the fake rate in simulation calculated through the TP fits and the true fake rate is then used as the third measure of the systematic uncertainty. The true fake rate is calculated by requiring the probe to match a generator-level electron in the identical way used to form the signal template sample. The systematic uncertainty obtained this way is less than 2% in most of the TP bins.

All three estimates are then combined in quadrature with equal weights in each TP bin, resulting in 2–5% systematic uncertainties for all bins. The statistical uncertainties on the fake rate values, on the other hand, vary from 1% in the most populated TP bins to 10% in the least.

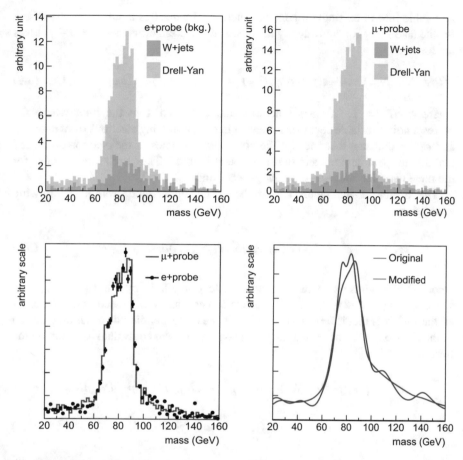

Fig. 5.4 *Top*: Tag-probe (*left*) and muon-probe (*right*) invariant mass distributions in simulation. Only the tag-probe events where the probe does not match an electron from $Z \to ee$ are used for the *left plot*. The contributions from W+jets (*yellow*) and DY (*green*) samples are stacked. *Bottom left*: Comparison of the overall shapes from the *two top panels*. *Bottom right*: An example of reshaping of the background in data. The *blue curve* (original muon-probe distribution) is multiplied by the ratio of the two distributions in the *bottom-left plot* to obtain the *red curve*. The wiggles in the *red curve* are due to statistical fluctuations

5.2.5   *Fake Rate Function Determination (eγ)*

Once the fake rate is measured in each TP bin, its parametrization with a function can be determined. As mentioned in Sect. 5.2.2, the fake rate is expressed as a function of the three variables p_T^e, N_{trk}, and N_{vtx}. This function is assumed to be separable, i.e., the electron selection efficiency of the inverted electron veto is factorable into three single-argument functions. Under this assumption, the ansatz for the fake rate f, or equivalently the electron selection efficiency $\epsilon = 1 - f$, takes the form

$$f(p_T^e, N_{\text{trk}}, N_{\text{vtx}}) = 1 - \epsilon(p_T^e, N_{\text{trk}}, N_{\text{vtx}}) = 1 - \epsilon_0 \cdot \epsilon_1(p_T^e) \cdot \epsilon_2(N_{\text{trk}}) \cdot \epsilon_3(N_{\text{vtx}}), \qquad (5.2)$$

where ϵ_i $(i = 1, 2, 3)$ are functions that approach 1 in the limit where the corresponding variable becomes irrelevant in discriminating electrons from photons, and ϵ_0 is a constant that encodes the other dependencies of the electron efficiency that are not parametrized and thus have been ignored. The "irrelevancy limits" for the three variables are $p_T^e \to \infty$, $N_{\text{trk}} \to \infty$, and $N_{\text{vtx}} \to 0$.

The origin of the factor ϵ_0 can be better understood from the following expression:

$$\epsilon^S = \int \epsilon(x_1, x_2, \ldots) \rho^S(x_1, x_2, \ldots) dx_1 dx_2 \ldots \qquad (5.3)$$

where ϵ^S represents the inclusive electron selection efficiency over the entire phase space of sample S, $\epsilon(x_1, x_2, \ldots)$ is the electron efficiency expressed as a function of all the variables that it depends on, and ρ^S is the corresponding distribution function of the electron in S. The separability ansatz above is then equivalent to assuming the form

$$\epsilon^S = \int \epsilon_1(p_T^e) \epsilon_2(N_{\text{trk}}) \epsilon_3(N_{\text{vtx}}) \rho^S(p_T^e, N_{\text{trk}}, N_{\text{vtx}}) dp_T^e dN_{\text{trk}} dN_{\text{vtx}}$$

$$\cdot \int \epsilon_r(y_1, y_2, \ldots) \rho_r^S(y_1, y_2, \ldots) dy_1 dy_2 \ldots, \qquad (5.4)$$

where ϵ_r and ρ_r^S are the remaining efficiency and distribution functions after the three dependencies are factored out. In other words, the electron efficiency and the probability distribution are both assumed to factorize into a function of the three variables of interest and the remainder. It follows that

$$\epsilon_0^S = \int \epsilon_r(y_1, y_2, \ldots) \rho_r^S(y_1, y_2, \ldots) dy_1 dy_2 \ldots, \qquad (5.5)$$

where ϵ_0^S is the constant factor mentioned above for the specific data set S.

From the above it becomes evident that the factorization in Eq. (5.2) is not exact, as, for instance, the fake rate is shown to have a dependency on the electron pseudorapidity η^e and the p_T^e, and η^e distributions are correlated. Furthermore, the constant factor ϵ_0^S is dependent on the sample used to calculate it. Therefore, the ansatz should be taken as a simplified approximation that will give a correct functional form up to certain errors, and its goodness is to be carefully evaluated later (see Sect. 5.2.9).

Even if the efficiency function is assumed to be factorable to three single-argument functions, Eq. (5.4) shows that the probability distribution is not necessarily separable. This situation can be simplified by noting that while p_T^e and N_{trk} are properties of the hard scattering, N_{vtx} is not, and thus the probability distribution should factorize into the form

$$p(p_T^e, N_{trk}, N_{vtx}) = \rho_H(p_T^e, N_{trk}) \cdot \rho_P(N_{vtx}). \qquad (5.6)$$

The final result is then

$$\epsilon = \epsilon_0' \int \epsilon_1(p_T^e)\epsilon_2(N_{trk})\rho_H(p_T^e, N_{trk})dp_T^e dN_{trk} \qquad (5.7)$$

$$= \epsilon_0'' \int \epsilon_3(N_{vtx})\rho_P(N_{vtx})dN_{vtx} \qquad (5.8)$$

where ϵ_0' and ϵ_0'' are constant factors that are obtained after all the variables except for the ones in the integrands are integrated out. The sample superscript S is omitted for simplicity. Note that ρ_H cannot be assumed to factorize, and therefore p_T^e or N_{trk} should not be integrated out to obtain a fake rate function dependent only on p_T^e or N_{trk}. In practical terms, Eqs. (5.7)–(5.8) imply that the fake rate must be determined as a 2-dimensional function in the (p_T^e, N_{trk}) plane, and can be independently parametrized as a 1-dimensional function of N_{vtx}.

The first step in the determination of the fake rate function is to study the MC truth fake rate to select the functional forms of ϵ_i that best parametrize the observed shapes. The following forms are chosen:

$$\epsilon_1(p_T^e) = 1 - \left(\frac{p_T^e}{c_1} + 1\right)^{-\alpha}, \qquad (5.9)$$

$$\epsilon_2(N_{trk}) = 1 - c_2 \cdot \exp\left(-\beta N_{trk}\right), \qquad (5.10)$$

$$\epsilon_3(N_{vtx}) = 1 - \gamma N_{vtx}, \qquad (5.11)$$

where $\alpha, \beta, \gamma, c_1, c_2$ are free positive floating parameters.

The values of the parameters in Eqs. (5.9)–(5.11) are fixed through least-χ^2 fits of the functions

$$1 - c'\epsilon_1(p_T^e; \alpha, c_1) \cdot \epsilon_2(N_{trk}; \beta, c_2)$$

and

$$1 - c''\epsilon_3(N_{vtx}; \gamma)$$

to the fake rates measured in each data TP bin. Here, c' and c'' are unimportant normalization factors. The surface and the line in the top left and top right panels of Fig. 5.3 represent the fit results. The residual difference of each point used in the fit, normalized by the corresponding uncertainty, is shown in the panels on the right-hand side of Fig. 5.3. The magnitude of the residual is less than the uncertainty value for most of the points.

The parameters obtained from the fits are

$$\alpha = 4.9 \pm 2.3,$$
$$\beta = 0.296 \pm 0.027,$$
$$\gamma = 0.000315 \pm 0.000012, \tag{5.12}$$
$$c_1 = 14 \pm 11 \, \text{GeV},$$
$$c_2 = 0.143 \pm 0.024.$$

The overall constant ϵ_0 in Eq. (5.2) still remains undetermined. The value of this factor is obtained from the self-consistency condition

$$n_{\text{fake}}^{\text{total}} = \sum_{i \in \text{all probes}} f(\mathbf{x}_i), \tag{5.13}$$

where $n_{\text{fake}}^{\text{total}}$ is the total number of fake (passing photon selection) probes, obtained by performing a TP fit to the entire numerator sample, and $\mathbf{x}_i$ represents the values of the variables of interest in each event. The sum on the right-hand side must be taken over all probes that are due to true electrons. However, obviously there is no way to tell the genuineness of the probe on a per-probe basis, which was in fact the reason for employing the TP technique to count electrons. A workaround is to use the set of electron proxy probes, which has a higher purity. Since a fraction $1/\epsilon$ of all the probes are proxy objects,

$$n_{\text{fake}}^{\text{total}} = \sum_{i \in \text{proxy}} \frac{f(\mathbf{x}_i)}{\epsilon(\mathbf{x}_i)} = \sum_{i \in \text{proxy}} \frac{1 - \epsilon_0 \cdot \epsilon_1(p_{\text{T}}^e{}_i) \cdot \epsilon_2(N_{\text{trk}i}) \cdot \epsilon_3(N_{\text{vtx}i})}{\epsilon_0 \cdot \epsilon_1(p_{\text{T}}^e{}_i) \cdot \epsilon_2(N_{\text{trk}i}) \cdot \epsilon_3(N_{\text{vtx}i})}. \tag{5.14}$$

The observed value of $\epsilon_0 = 0.9981$ leads to the final expression for the fake rate

$$f(p_{\text{T}}^e, N_{\text{trk}}, N_{\text{vtx}}) = 1 - 0.9981 \cdot \left[1 - \left(\frac{p_{\text{T}}^e}{14 \, \text{GeV}} + 1 \right)^{-4.9} \right]$$
$$\cdot [1 - 0.143 \cdot \exp(-0.296 \cdot N_{\text{trk}})] \cdot [1 - 0.000315 \cdot N_{\text{vtx}}]. \tag{5.15}$$

5.2.6   Systematic Uncertainty on the Fake Rate Function (eγ)

The uncertainty on the constant factor ϵ_0 is used to estimate the overall uncertainty of the fake rate function. The dominant uncertainty originates from the fit parameters, and is calculated by observing how much ϵ_0 changes when α, β, and γ from Eq. (5.12) are moved within their fit uncertainties.

The specific algorithm used to determine the systematic uncertainty on f utilizes the so-called pseudo-experiments as follows. The parameters α, β, and γ are randomly fixed to values pulled from normal distributions centered at the nominal values of the parameters. The widths of these distributions are set to the uncertainties in Eq. (5.12). The fake rate function is then refit to the measured fake rates to fix the remaining parameters c_1, c_2 of the function in Eqs. (5.9)–(5.10), and to recalculate the ϵ_0 factor using the new set of parameters. This procedure is repeated 1000 times to obtain a distribution of the ϵ_0 value. The 1σ interval of this distribution, 0.0002, is then taken as the systematic uncertainty on ϵ_0.

The uncertainty on ϵ_0 is propagated to the overall uncertainty on the fake rate function as

$$\frac{\delta f}{f} \sim \frac{\delta \epsilon_0}{f^{\text{total}}} = 14.2\%, \tag{5.16}$$

where f^{total} is the inclusive fake rate for all probes with $p_T{}^e > 25$ GeV.

5.2.7   Dependence of the Fake Rate on Track Multiplicity

The dependency of the electron-to-photon fake rate on N_{trk} arises from the implementation of the ECAL-driven electron reconstruction (see Sect. 3.4.3) that becomes inefficient in low-multiplicity events. In the ECAL-driven electron seeding, for each ECAL supercluster that passes the loose E_T and H/E cuts, trajectory seeds (see Sect. 3.4.1) that are consistent with the possible helix of the electron track are searched for. The seed finding tests two helices, one for each charge assumption, that connects the center of mass of the supercluster and the beam spot. The cluster E_T is used to calculate the curvature of the helices. The innermost hit of the trajectory seed must lie in a wide $z - \phi$ window around the point where the helix crosses the corresponding tracker layer. If the trajectory seed satisfies this condition, the z position of the primary vertex is re-approximated using the position of the innermost hit, and another hit in the trajectory seed that is consistent with the helix is searched for. If one such hit is found, the trajectory seed becomes the electron seed.

As mentioned in Sect. 3.4.1, the trajectory seeding proceeds in iterations. The algorithm of the first iteration selects triplet hits from the pixel tracker and has approximately 82% efficiency for electron tracks. The second iteration, which brings the overall efficiency to approximately 96%, selects pairs of pixel hits using

Fig. 5.5 Electron seeding
efficiency after each iteration
of trajectory seed
reconstruction in the DY
simulation

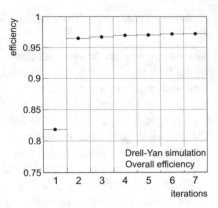

pixel vertices as the constraint in the track segment identification. Pixel vertices are primitive vertices formed using only the pixel tracker hit information before complete tracks are reconstructed. Being limited by the available information, the pixel-vertexing algorithm is not capable of recognizing the hard-scattering vertex if there are too few tracks emerging from it. Thus the second trajectory-seeding iteration is inefficient when the charged-particle multiplicity of the primary scattering is low. Since the true number of charged particles cannot be known in data, N_{trk} is used as a proxy to this quantity.

This point is confirmed in the DY MC simulation data set by considering the trajectory seeds formed in each iteration separately, and for each iteration measuring the efficiency of successfully forming a seed for an electron. Figure 5.5 shows the overall electron seeding efficiency after each iteration of the trajectory seeding. There are seven iterations in the trajectory seed search for the electrons. Figure 5.6 then breaks down the efficiencies in bins of charged-particle multiplicity of the hard scattering. A dependency on the latter is clearly observed after the second iteration, and the inefficiency in bins with low number of tracks is not covered in later iterations.

5.2.8 $\mu\gamma$ *Channel*

The electron proxy sample in the $\mu\gamma$ channel is defined in a similar way to the $e\gamma$ channel; the photon candidate object is replaced by an electron-veto-inverted proxy object. Similarly to the $e\gamma$ channel, when multiple electron proxy objects are identified in an event, each proxy-muon pair is used in the background estimation independently.

The fake rate in the $\mu\gamma$ channel is different from that in the $e\gamma$ channel, due to the different trigger constraints. In particular, the tracker isolation present in the diphoton trigger lowers the fake rate in the $e\gamma$ channel. Since electron-to-photon fakes are a minor background in the $\mu\gamma$ channel, instead of re-deriving the full fake rate function, the functional forms and parameter values of $\epsilon_1(p_T^e)$, $\epsilon_2(N_{trk})$, and $\epsilon_3(N_{vtx})$ are unchanged, and only the ϵ_0 factor is re-calculated for the $\mu\gamma$ channel.

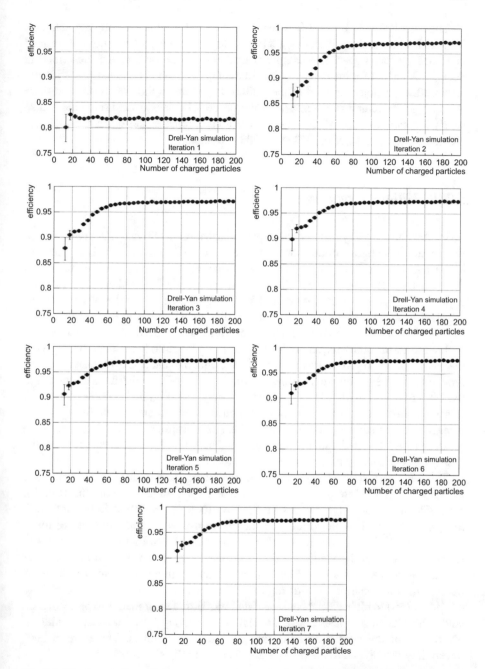

Fig. 5.6 Electron seeding efficiency as a function of the charged-particle multiplicity of the hard scattering, after each iteration of trajectory seed reconstruction in the DY simulation sample

The calculation in Eq. (5.14) is repeated with the SingleElectron data set. This data set is chosen since it allows a TP sample where the probe is not constrained or biased by the trigger. The sample preparation is identical to that described in Sect. 5.2.3, except for the trigger object matching. The tag is matched to the single-electron trigger object instead of the trailing leg of the diphoton trigger, and there is no trigger-matching requirement for the probe. The background template is also nearly identical to Sect. 5.2.3, but there are no additional isolation requirements to the probe. An ϵ_0 value of 0.0023 ± 0.0003 is obtained.

5.2.9 Validity Evaluation

To evaluate the validity of the chosen parametrization of the fake rate, two tests on the electron fake background estimation are performed. The first test is simulation-based and assesses the applicability of the parametrized fake rate to processes different from $Z \rightarrow ee$. The $e\gamma$ channel and $\mu\gamma$ channel candidate selections are applied to a mixture of the DY, $t\bar{t}$, and WW MC samples, with the photon objects matched to the generator-level electrons. This target sample is then compared to the weighted electron proxy sample taken from the same data sets, in this case without truth-matching requirement. The fake rate function is obtained by applying the parametrization procedure described in Sect. 5.2.5 to the MC truth fake rate. This type of tests where a quantity estimated fully within MC simulation is compared to the MC truth information is called a MC closure test. Figure 5.7 shows the result of such a closure test. The estimation method is shown to predict the full spectrum of each variable studied within uncertainties. The discrepancy of 1% between the estimated and counted number of electron fakes is added in quadrature to the systematic uncertainty of the fake rate used for real data.

The second test used real data and evaluated how well the electron fake estimation can be extrapolated to the signal region in E_T^{miss}. For this test, the SingleElectron data set is binned in E_T^{miss} up to $E_T^{\mathrm{miss}} < 70\,\mathrm{GeV}$. In each E_T^{miss} bin, the TP method is performed on the electron-plus-photon events to estimate the number of Z-decay electrons faking photons. This estimation is compared to the prediction given by weighting the electron-plus-proxy events with the transfer factor. Figure 5.8 shows the estimated and predicted number of electron pairs. A trend of underestimation towards high E_T^{miss} is observed. While it is not possible to extend this test to higher E_T^{miss} because data lacks sufficient statistics to perform a TP study, the MC closure tests above suggest that the trend does not cause significant divergence of the prediction from the true number of fakes. Therefore, a 20% uncertainty is assigned to the proxy sample shape in the signal region.

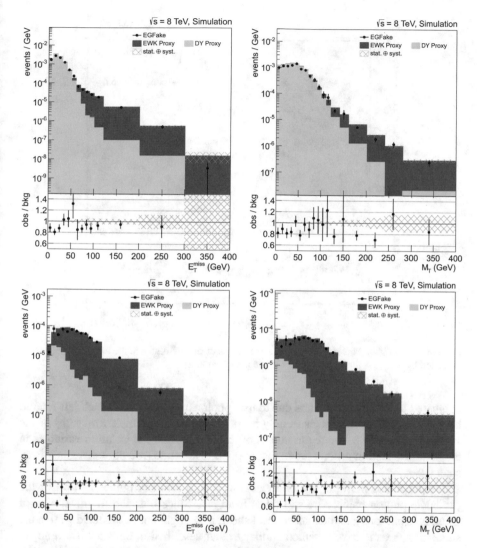

Fig. 5.7 MC closure test for electron fake estimation in the $e\gamma$ (*top*) and $\mu\gamma$ (*bottom*) channels. The fake rate determined using the Drell–Yan sample is applied to the mixture of the DY, $t\bar{t}$, and WW samples. In the figure, EWK refers to the combination of the latter two samples. Closure tests in the distributions of E_T^{miss} (*left*) and M_T (*right*) are shown

5.3 Fake Photon Background Due to Jets

5.3.1 Overview

Electromagnetically fluctuating jets, such as those whose energy is mostly carried by a single π^0 that decays to two photons, can mimic photon signals. In general,

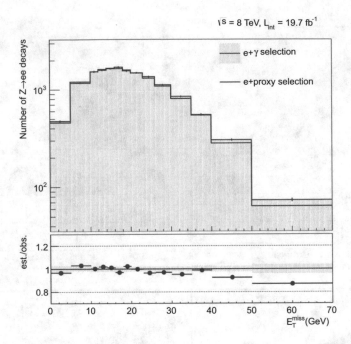

Fig. 5.8 Comparison between the TP-estimated number of electrons and the number of electrons predicted by applying the transfer factor to the proxy sample

compared to the ECAL clusters due to true photons, the ones from such jets tend to have wider shower shapes, especially in the η direction, and higher p_T sums of the surrounding objects. The requirements on H/E, $\sigma_{i\eta i\eta}$, and the isolation variables in the photon selection criteria are in fact applied to separate true-photon signals from such fakes.

Despite these multiple layers of jet rejection, a non-negligible fraction of the candidate photon objects are still fakes from jets. Since both the fragmentation of hadrons to photons and their subsequent showering might not be modeled well in the simulation, a data-driven method similar to that described in Sect. 5.2 is employed to estimate the contribution from such background events into the signal region. The method works by scaling the hadron proxy sample, which is a sample whose event content is a photon-like jet plus lepton, with the estimated ratio of the number of fake photons to the number of proxy objects (transfer factor). The photon-like jet is called the hadron proxy object in the following. The transfer factor is derived from the estimation of the fraction of hadronic fakes within the candidate photon sample. The isolation condition in the diphoton HLT for the $e\gamma$ channel, which uses different variables from the offline isolation and thus effectively acts as another set of isolation cuts, lowers the fake fraction in the $e\gamma$ channel. Additionally, the hadron proxy objects are taken from the sample of photon objects that match the respective HLT objects in the two search channels. Therefore, the background estimation is performed separately for the $e\gamma$ and $\mu\gamma$ channels.

It should be noted that a significant fraction of the hadron proxy sample can be events with fake leptons. Consequently, by scaling this sample with the transfer factor, an estimate of the double-fake background events is given simultaneously together with that for the jet-to-photon background.

In the following, the $\mu\gamma$ channel background estimation is first described in detail. The estimation is done in multiple steps. The first step is to determine the fraction of hadronic fakes within the candidate photons in a sample where the photon objects are free of trigger constraints. This fraction is subsequently applied to the sample of candidate photons in a control region to construct the estimation of the p_T distribution of the hadronic fakes. The ratio of this p_T distribution to the distribution of the hadron proxy objects will be the transfer factor as a function of proxy object p_T. Finally, this function is used as a weight to the hadron proxy sample.

5.3.2 The Hadron Fraction ($\mu\gamma$)

The fraction of fake photons, or the hadron fraction, within the candidate sample is inferred from a template fit to the $\sigma_{i\eta i\eta}$ distribution of the photon objects. As stated in Sect. 5.3.1, fake photons tend to have laterally wider shower shapes than true photons. It follows that if it were somehow possible to gather pure fake and true photon samples independently, the $\sigma_{i\eta i\eta}$ distribution of the former would be wider than the latter. Accordingly, if the candidate sample dominantly consists of hadronic fakes and true photons, it should be possible to describe the $\sigma_{i\eta i\eta}$ distribution of the candidate photons as a linear combination of the hadronic distribution and the photonic distribution.

The measurement of the hadron fraction within the candidate photons is performed on events from the SingleMu data set. The events are selected with a requirement of $E_T^{\text{miss}} < 70\,\text{GeV}$ and at least one muon object, defined in Sect. 4.5, that matches the single-muon trigger object. If there are multiple such muons, the one with the highest p_T is used. Within such events, photon objects with $E_T > 25\,\text{GeV}$ are looked for. The photon object must also satisfy the conditions on H/E, I_{NH}, I_{Ph}, and the electron vetoes defined in Sect. 4.3. The events must also pass the selection criteria for the FSR rejection given in Sect. 4.6. This sample of photons is then binned finely in E_T^γ. Within each E_T^γ bin, a two-component template fit is performed to identify a linear combination of the hadronic and photonic $\sigma_{i\eta i\eta}$ distributions that best describes the distribution of the candidate photons, as mentioned in the previous paragraph. An example of such a template fit is displayed in Fig. 5.9. The target distribution of the fit, shown with black dots in Fig. 5.9, is the $\sigma_{i\eta i\eta}$ distribution of the photon objects additionally passing the selection criterion on I_{CH}, i.e., those passing all the photon selection criteria except for the one on $\sigma_{i\eta i\eta}$ ("$N-1$" objects). The same distribution of the photon objects failing the selection on I_{CH}, but with an upper bound of $I_{\text{CH}} < 15\,\text{GeV}$, constitutes the hadronic template, which corresponds to the red histogram in the figure. The blue histogram, stacked on top of the red, is the photonic template, which is described below. The two templates

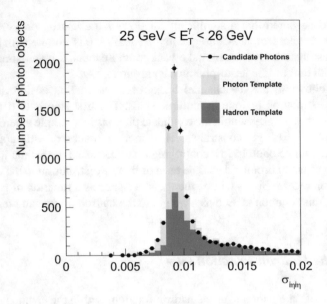

Fig. 5.9 The result of the $\sigma_{i\eta i\eta}$ template fit in $E_T{}^\gamma$ bin $25\,\text{GeV} < E_T{}^\gamma < 26\,\text{GeV}$

are assigned floating scale factors, whose values are determined through a least-χ^2 fit using each $\sigma_{i\eta i\eta}$ bin in Fig. 5.9 as a data point. The statistical uncertainties of the template and target histograms, where the ones of the templates are multiplied by the respective scale factors, are included in the denominator of χ^2.

To form the photonic templates corresponding to the blue histogram in Fig. 5.9, the $Z \rightarrow \mu\mu\gamma$ FSR is exploited as the source of a high-purity photon sample. The events are again collected from the single-muon triggered sample, but this time two muons are required in each event. One of the muons must have $p_T > 25\,\text{GeV}$ and pass the muon selection criteria without the isolation requirement. The other muon is only required to be a PF-reconstructed, global or tracker muon. The invariant mass of the dimuon system must be less than 80 GeV. Within the dimuon events, a photon candidate with $E_T > 25\,\text{GeV}$ passing the loose criteria without the condition on $\sigma_{i\eta i\eta}$ is looked for. All such photons that are found within $\Delta R < 0.8$ of one of the muons and with a $\mu\mu\gamma$ three-body invariant mass between 81 and 101 GeV are used for the photonic template. The two cuts on the invariant masses as well as the criterion on the proximity of the photon and the muons reduce the contribution of the photon ISR events, which can be seen as a photon radiation largely uncorrelated with the production of the Z boson. In such events, the purity of the photon object cannot be guaranteed since it is possible that the photon object is due to an electromagnetic jet (Z+jets). On the other hand, muons do not radiate gluons, and thus the photon objects in the events with the invariant mass of the $\mu\mu\gamma$ system close to m_Z can be expected to have high purity. Since FSR photons have a steeply falling E_T distribution, the $E_T{}^\gamma$ bins corresponding to photons with E_T higher than 50 GeV suffer from poor statistics. To work around this problem, a fixed photonic template

Fig. 5.10 Hadron fractions
in $\mu\gamma$ channel

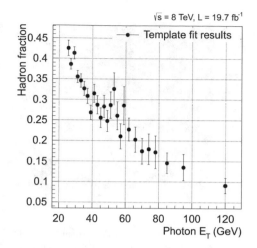

for those bins is used, as the $\sigma_{i\eta i\eta}$ distribution of true photons is known to become insensitive to E_T^γ above $\sim$40 GeV. This fixed template is formed by all photons with $E_T^\gamma > 50$ GeV.

Once the normalization of the hadronic template is determined, the hadronic template and the target distributions are integrated over the photon selection range of $0 < \sigma_{i\eta i\eta} < 0.012$. The ratio of these integrals gives the estimate of the fraction of hadrons within the set of photon candidates in each E_T^γ bin. The measured fraction plotted against E_T^γ is shown in Fig. 5.10. The E_T^γ coordinates of the points are simply the centers of the bins used for the template fits, and the errors on the fraction values represent the uncertainty of the fit.

The estimated hadron fractions vary rather sensitively by changes in the fit template shapes. However, it is known that there is a slight correlation between I_{CH} and $\sigma_{i\eta i\eta}$, due to the inclusion into the photon cluster of the electromagnetic energy deposits by the charged hadrons surrounding the photon object. Thus certain errors on the hadron fractions are to be expected because the I_{CH} values of the events in the hadronic template differ from those of the fake events by construction. The size of this uncertainty is assessed by modifying the hadronic template definitions and repeating the hadron fraction measurements. Figure 5.11 shows the extracted hadron fraction in one of the E_T^γ bins as a function of the upper and lower cut on I_{CH} of the hadronic template. Similar scans are performed for each E_T^γ bin, and the standard deviation of the fractions within the scan range is used as the uncertainty on the fraction for each E_T^γ bin. This systematic uncertainty on the hadron fraction, on the order of 1–2%, is added in quadrature to the fit uncertainty.

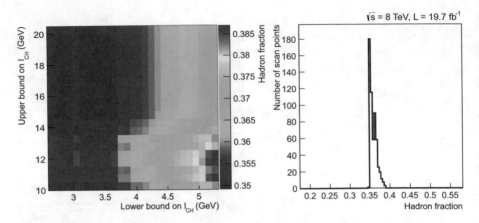

Fig. 5.11 *Left*: Extracted hadron fractions in bin $32\,\text{GeV} < E_T^\gamma < 34\,\text{GeV}$ as a function of the upper (vertical axis) and lower (horizontal axis) I_{CH} cuts for the hadronic template. *Right*: Distribution of the fractions in the range examined in the *left plot*

5.3.3 Proxy Sample and Transfer Factor ($\mu\gamma$)

The hadron fraction determined above does not strongly depend on the details of the event selection. Therefore, the E_T^γ distribution of the muon candidate sample in the control region $E_T^{\text{miss}} < 70\,\text{GeV}$ is multiplied by the hadron fractions, and the result is used as the estimate of the p_T distribution of the hadronic fakes. Since it is useful to have the transfer factor in a parametrized form, the p_T distributions of the fakes and proxy objects are described by simple analytic functions, and the ratio of these functions is used as the transfer factor. The denominator, which expresses the p_T distribution of the hadron proxy objects, is determined first.

The hadron proxy sample is collected from the MuEG data set, with the muon selection identical to the signal candidate muon. From this sample, instead of the candidate photons, barrel photon objects with $E_T^\gamma > 25\,\text{GeV}$ that pass all photon selection criteria except for those on the $\sigma_{i\eta i\eta}$ or I_{CH} are collected. The photons that fail either one or both of the requirements for $\sigma_{i\eta i\eta}$ and I_{CH} are used as hadron proxy objects. The trigger object matching is also required for the proxy objects, which constrains the range of $\sigma_{i\eta i\eta}$ value to $\sigma_{i\eta i\eta} \lesssim 0.014$. The upper bound is not strict because of the calibration differences between the HLT and offline reconstructions. Additionally, I_{CH} is bounded from above at $15\,\text{GeV}$ to keep the kinematic properties of the hadron proxy objects to be not too different from the hadronic fakes. If multiple hadron proxy objects are found in an event, each muon-proxy combination is regarded as one proxy event.

The proxy p_T distribution is obtained from the proxy events with $E_T^{\text{miss}} < 70\,\text{GeV}$. The p_T distribution is parametrized by a sum of two exponentials, and the parameter values are determined with an unbinned maximum-likelihood fit. The left-hand plot of Fig. 5.12 shows the distribution and fit function. The 1σ uncertainty band around

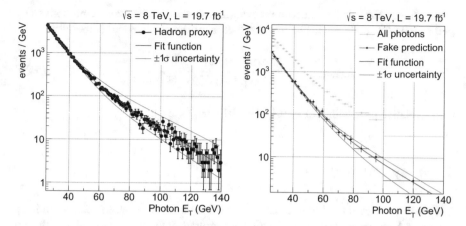

Fig. 5.12 *Left*: p_T distribution of the hadron proxy objects in $\mu\gamma$ channel. *Right*: Estimated p_T distribution of the fake photons in $\mu\gamma$ channel

the fit function is obtained by recalculating the fit function repeatedly with the function parameters modified about their nominal values. These modified values of the parameters were drawn from normal distributions with the spreads given by the respective uncertainties of the parameters from the original fit. The bands are constructed from the envelope of 100 such modified functions. It is observed that the relative uncertainty increases approximately linearly with respect to p_T. The uncertainty is less than 10% at the low end of the p_T spectrum, and reaches 100% around a p_T value of 150 GeV.

The numerator function is derived by minimizing a χ^2 value defined as

$$\chi^2 := \sum_j \frac{\left(\int_j dp_T F(p_T) - n_j\right)^2}{\sigma_j^2}, \tag{5.17}$$

where j runs over the bin numbers and $\int_j dp_T F(p_T)$, n_j, and σ_j are, respectively, the integral of the trial function within the p_T bin j, the bin content, and its uncertainty. The sum of two exponentials is again used as the trial function. The estimated fake p_T distribution together with the result of the fit is shown in the right-hand plot of Fig. 5.12. The uncertainty band on the fit function is obtained by repeating the pseudo-experiment method described above.

From the two fits, the distribution functions

$$F_{\text{proxy}}(p_T) = (7.9 \times 10^4) \cdot \exp\left(-0.120 \cdot \frac{p_T}{\text{GeV}}\right) + (1.9 \times 10^3) \cdot \exp\left(-0.048 \cdot \frac{p_T}{\text{GeV}}\right) \tag{5.18}$$

and

$$F_{\text{fake}}(p_T) = (5.1 \times 10^4) \cdot \exp\left(-0.118 \cdot \frac{p_T}{\text{GeV}}\right) + (7.4 \times 10^2) \cdot \exp\left(-0.048 \cdot \frac{p_T}{\text{GeV}}\right) \tag{5.19}$$

are obtained. The relative uncertainties on the two function values are both given by $(-0.1 + 0.008 \cdot p_T/\text{GeV})$. The transfer factor for the $\mu\gamma$ channel is therefore

$$R_{\text{fake/proxy}} = \frac{27 \exp\left(-0.070 \cdot \frac{p_T}{\text{GeV}}\right) + 0.39}{42 \exp\left(-0.072 \cdot \frac{p_T}{\text{GeV}}\right) + 1}, \tag{5.20}$$

with a relative uncertainty of $(-0.14 + 0.011 \cdot p_T/\text{GeV})$.

5.3.4   eγ Channel

The difference in the trigger constraint on the photons between the eγ and $\mu\gamma$ channels prompts a slightly different background estimation method for the eγ channel compared to the $\mu\gamma$ channel discussed above. The hadron fake fraction is still determined using the SingleMu data set, but the targets of the $\sigma_{i\eta i\eta}$ template fits are formed from photons with offline isolation requirements $I_{\text{ECAL}} < 5\,\text{GeV}$, $I_{\text{HCAL}} < 5\,\text{GeV}$, and $I_{\text{trk}} < 5\,\text{GeV}$ on top of the selection criteria applied for the $\mu\gamma$ channel measurement.

As is the case in the $\mu\gamma$ channel, the numerator of the transfer factor is obtained from the p_T distribution of the hadronic fakes in the control region $E_T^{\text{miss}} < 70\,\text{GeV}$. In the e$\gamma$ channel, however, there is a large contribution from electron-to-photon fakes in the candidate photon sample, due to the relatively high cross section of the DY dielectron process. Therefore, this component is first subtracted before applying the hadron fraction:

$$n_j^{\text{hadron}} = h_j(N_j^{\text{cand}} - n_j^e), \tag{5.21}$$

where n_j^{hadron} is the estimated number of hadronic fakes, h_j the hadron fraction, N_j^{cand} the observed number of candidate photons, and n_j^e is the estimated number of electrons, all in the jth E_T^γ bin. The number of electrons n_j^e is estimated using the formula Eq. (5.14) in each E_T^γ bin.

The hadron proxy sample in the eγ channel is taken from the Photon/DoublePhoton data set, requiring a candidate electron and a hadron proxy object, which is defined identically to Sect. 5.3.3, i.e., pass all photon selections except for the ones on $\sigma_{i\eta i\eta}$ or I_{CH}. A match to the leading leg of the diphoton trigger is required for the proxy. Such proxy objects are not purely hadronic but also contain electrons, despite the inverted cuts on isolation and shower shape. Nevertheless, no subtraction is performed for this contribution, since the electrons will equally exist in the proxy sample used in the actual background estimation. In other words, the transfer factor is considered to be the ratio between the number of hadrons in the signal region to the number of all objects in the proxy sample, allowing some non-hadronic contaminations. As long as the contamination does not significantly alter the kinematic distributions of the proxy sample to be too dissimilar to the hadronic fakes, the background estimation will be valid within uncertainties.

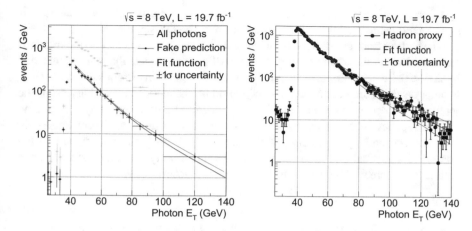

Fig. 5.13 *Left*: Estimated p_T distribution of the fake photons in the $e\gamma$ channel. *Right*: p_T distribution of the hadron proxy objects in the $e\gamma$ channel

Figure 5.13 shows the p_T distributions of the hadronic fakes and their proxy objects in the $e\gamma$ channel. From the fits,

$$F_{\text{fake}}(p_T) = (1.6 \times 10^4) \cdot \exp\left(-0.095 \cdot \frac{p_T}{\text{GeV}}\right) + (4.4 \times 10^2) \cdot \exp\left(-0.043 \cdot \frac{p_T}{\text{GeV}}\right)$$

(5.22)

and

$$F_{\text{proxy}}(p_T) = (7.3 \times 10^4) \cdot \exp\left(-0.105 \cdot \frac{p_T}{\text{GeV}}\right) + (2.4 \times 10^3) \cdot \exp\left(-0.043 \cdot \frac{p_T}{\text{GeV}}\right)$$

(5.23)

are obtained, where again the relative uncertainties on the function values are $(-0.1 + 0.008 \cdot p_T/\text{GeV})$. The transfer factor for the $e\gamma$ channel is therefore

$$R_{\text{fake/proxy}} = \frac{6.8 \exp\left(-0.052 \cdot \frac{p_T}{\text{GeV}}\right) + 0.19}{30 \exp\left(-0.062 \cdot \frac{p_T}{\text{GeV}}\right) + 1},$$

(5.24)

with a relative uncertainty of $(-0.14 + 0.011 \cdot \frac{p_T}{\text{GeV}})$.

5.3.5 Closure Test

The validity of the jet-to-photon fake estimation method is again assessed with a MC closure test. The test is performed on a mixture of the W+jets and DY samples. Similarly to the measurement on real data, the hadron fraction within each E_T^γ bin is determined by a $\sigma_{i\eta i\eta}$ template fit. The target and hadronic templates are obtained by the same procedure as described in Sect. 5.3.2, whereas the photonic template is constructed using the MC truth information from the DY sample.

The hadron fractions derived from template fits are then compared to the fractions calculated using the MC truth information. All photon objects that are not matched by $\Delta R < 0.1$ to a generator-level prompt photon or electron are considered as fakes. Prompt particles are defined as those which do not have mesons and baryons as the mother in the decay chain. The level of non-closure is then expressed by normalizing the discrepancy in the fractions by their inherent uncertainties. Discrepancies for both the $\mu\gamma$ and $e\gamma$ channels are approximately 30% greater compared to the combined statistical and fit uncertainties. The uncertainty on the transfer factor in data is thus increased by the same relative amount accordingly.

5.4   Background Due to Lepton Mis-identification

5.4.1   Overview

The signal candidate events contain not only events with fake photons but also those with fake leptons. All reconstructed lepton objects that do not directly originate from decays of W or Z bosons are considered as fakes. Thus leptons from heavy-flavor decays are also fakes in this definition. In fact, a study of simulated QCD events shows that most of the fake muons are products of B meson decays. On the other hand, fake electrons are found to be mostly due to light-flavor jets.

Similar to the determination of the fake photon backgrounds, a proxy sample is constructed to estimate the contribution of fake leptons in the signal region. Each event in the proxy sample must have at least one candidate photon, one fake lepton proxy, and no candidate lepton. This sample is scaled with a normalization factor to give the background estimation in the various observables. The normalization factor for the lepton fakes is derived through template fits using the quantity $\Delta\phi(\ell, E_T^{\mathrm{miss}})$ in a control region together with the Vγ background, as discussed in Sect. 5.5.2.

The proxy sample contains not only events with a true photon and a QCD jet, but also events where the photon object is a mis-identified hadron (the double-fake background), in a similar manner to the fake photon proxy sample containing fake lepton events. Such events will arise dominantly from QCD dijet processes, whereas those with a true photon would come from γ+jet process. However, the two processes are kinematically similar, and thus the proxy sample defined above can represent either process. Since the double-fake background is mostly accounted for by the jet-to-photon fake estimation, the proxy sample above should be normalized to only cover the fake lepton background.

5.4.2 Fake Electron Proxies

By studying reconstructed electron objects in the QCD simulation, it can be shown that fake electrons are typically isolated light-flavor hadrons that shower significantly in the ECAL. One characteristic of the electron object in such a case is that its track momentum and the cluster energy are less consistent with each other compared to a genuine electron. This feature is exploited in the following definition of the fake electron proxy object:

- Pass selection criteria on $\sigma_{i\eta i\eta}$, H/E, $|d_0|$, $|d_z|$, and $|1/E - 1/P|$. Pass conversion veto.
- Fail either one or both of the selection criteria on $|\Delta\eta_{\text{in}}|$ and $|\Delta\phi_{\text{in}}|$.
- Total isolation sum must be greater than $10\,\text{GeV}$.

When applied in simulation, this selection is shown to have an efficiency on prompt electrons that is less than $1/1000$ of the efficiency of the electron candidate selection. The efficiency on hadrons and non-prompt leptons, on the other hand, is higher than that of the candidate selection.

The kinematic distributions of the events with fake and proxy electrons are compared in the simulation to check the validity of the background modeling. Ideally, a QCD prompt photon simulation should be used for this study, where the fake electron events are collected with the full analysis selection, and the proxy sample contains the candidate photon object in addition to the fake electron proxy. However, due to constraints from the data set size, the dijet QCD MC samples are used, and the event selection is tuned accordingly. Instead of a photon, events are required to have at least one jet with $p_T > 40\,\text{GeV}$ and $|\eta| < 1.5$ that is separated from the fake or proxy electron object by $\Delta R > 0.8$. Figure 5.14 shows the result of the comparisons. The variables $\Delta\phi(\ell, E_T^{\text{miss}})$, E_T^{miss}, M_T, and H_T are studied. The first variable $\Delta\phi(\ell, E_T^{\text{miss}})$ is plotted with a $E_T^{\text{miss}} > 20\,\text{GeV}$ selection, since this variable is used in the normalization determination in Sect. 5.5.2 with a E_T^{miss} cut. The distributions mostly agree within statistical uncertainties. The discrepancy in the tail of the M_T distribution is addressed in Sect. 6.3 as shape uncertainty of the proxy sample.

5.4.3 Fake Muon Proxies

As already mentioned, unlike fake electrons, the simulation shows that the fake muons are mainly non-prompt true muons from heavy-flavor decays. This implies that there is little handle to suppress the contribution from prompt muons to the proxy sample, since there is no inconsistency within the reconstructed object to exploit. Therefore, the definition of the fake muon proxy object solely relies on the isolation variable, i.e., it must pass all of the muon candidate selection criteria except for that on the relative isolation, which must have a value between 0.15 and 0.6.

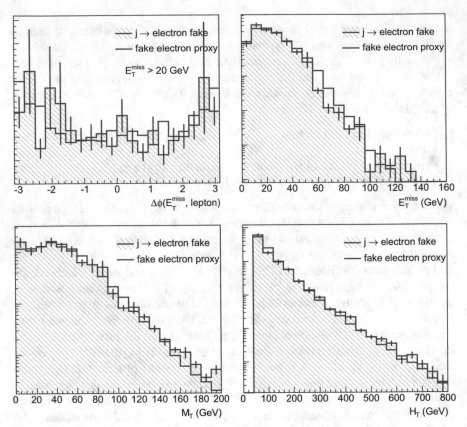

Fig. 5.14 *Top left to bottom right*: Comparisons of $\Delta\phi(\ell, E_T^{miss})$, E_T^{miss}, M_T, and H_T distributions between dijet simulation events with a fake electron and those with a proxy electron. The distributions are normalized to have equal area. A $E_T^{miss} > 20\,\text{GeV}$ requirement is applied for the $\Delta\phi(\ell, E_T^{miss})$ distribution

This proxy object definition has an efficiency of approximately 5% for selecting prompt muons. While this value is significantly higher than in the eγ channel, the fraction of prompt muons within the proxy sample is still found to be sufficiently low.

The validity of the background modeling is again checked on the dijet QCD simulation sample. The requirement of a photon candidate is replaced by a jet in the same way as in the eγ channel. Figure 5.15 shows the result of the comparisons on the same variables studied for the fake electrons. Due to the limited data set size, the E_T^{miss} cut could not be applied for the $\Delta\phi(\ell, E_T^{miss})$ distribution. The fake and proxy distributions again agree within statistical uncertainties.

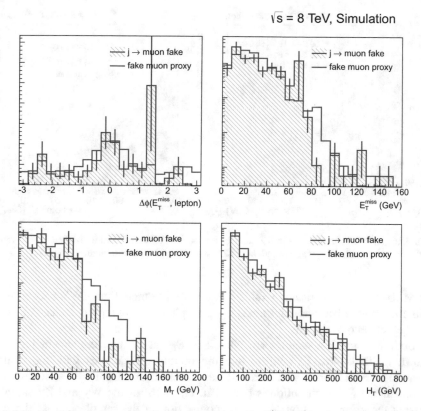

Fig. 5.15 *Top left to bottom right*: Comparisons of $\Delta\phi(\ell, E_T^{miss})$, E_T^{miss}, M_T, and H_T distributions between dijet simulation events with a fake muon and those with a proxy muon. The distributions are normalized to have equal area

5.5 EWK Background

5.5.1 The Standard Model Vγ Cross Section

The standard model production of a W or Z boson accompanied by a photon (Vγ production) is the main background of this search. Currently, neither the CMS nor the ATLAS collaboration has published cross section measurements of these processes at $\sqrt{s} = 8$ TeV. While the cross section is calculable up to NLO in QCD with existing tools such as MCFM [2], it was seen in the $\sqrt{s} = 7$ TeV measurements that the calculation of the Wγ cross section might underestimate the observed value significantly [3, 4]. On the other hand, the matrix-element-based event generation with extra partons using SHERPA [5] or MADGRAPH was found to reproduce the key kinematic features of the Vγ events well. Therefore, Wγ and Zγ data sets generated

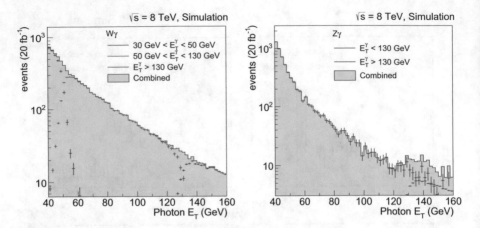

Fig. 5.16 $E_T{}^\gamma$ distributions in the $\mu\gamma$ channel normalized to $20\,\text{fb}^{-1}$, in individual Wγ samples and their combination (*left*) and in individual Zγ samples and their combination (*right*)

by MADGRAPH 5, as listed in Table 4.6, are used to model the EWK background, with their normalizations determined from data in a control region, instead of using the theoretical calculations.

Both Wγ and Zγ samples are binned in $E_T{}^\gamma$ at the matrix-element level, as shown in Table 4.6. To confirm that the mixture of samples with the weight given in Eq. (4.1) forms a background sample valid for the full $E_T{}^\gamma$ spectrum, the reconstructed $E_T{}^\gamma$ distribution is plotted in Fig. 5.16 for the Wγ and Zγ samples passing the event selection. A smooth connection of the samples across the $E_T{}^\gamma$ bins can be seen. The plots are normalized to a data size of $20\,\text{fb}^{-1}$ to give a rough estimate of the number of Vγ events that are expected in the signal candidate sample. The Wγ and Zγ samples are then further combined into a single Vγ sample, with the relative weights used in Fig. 5.16. The absolute normalization of the Vγ sample is determined by the method described below.

5.5.2 Normalization of the Vγ and Fake Lepton Samples

Since there are two background categories, i.e., Vγ and fake leptons, that are not given absolute normalizations, the size of their contributions to the signal region is determined simultaneously. Template fits in the $40\,\text{GeV} < E_T^{\text{miss}} < 70\,\text{GeV}$ control region are employed for this purpose. Templates taken from the two samples are scaled to form a linear combination that best describes the distribution of the signal candidate sample. The best-fit coefficients are then taken as the scale factors applied to the samples. This fit procedure is performed in the $e\gamma$ channel and $\mu\gamma$ channel independently. Since the resulting scale factors on the Vγ templates are essentially correction scale factors to the calculated cross sections of the V$\gamma \to \ell\nu\gamma$ processes, the value obtained in the two channels is expected to agree.

Fig. 5.17 E_T^{miss} distributions of the Wγ and Zγ samples, normalized to 20 fb^{-1}

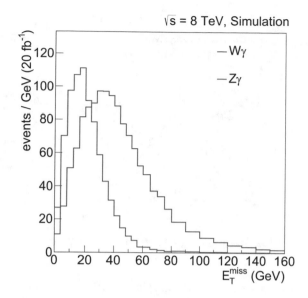

The lower cut on E_T^{miss} at 40 GeV in the definition of the control region is applied to reduce the Zγ component in the Vγ background. Figure 5.17 shows the E_T^{miss} distributions of the Wγ and Zγ samples, again normalized to 20 fb^{-1}. It is observed that a large fraction of Zγ events will be rejected with the cut, mitigating the uncertainty related to assigning a single scale factor to the Vγ background, which is in fact a composite sample.

The distributions of $\Delta\phi(\ell, E_T^{miss})$ are used as templates. First, estimated fake photon and rare EWK background (see Sect. 5.5.3) components are subtracted from the observed $\Delta\phi(\ell, E_T^{miss})$ distribution in data. The resulting histogram, called the target, is then fit by two templates, one from the Vγ sample and the other from the fake lepton proxy sample, with floating scales. The scales are determined by a least-χ^2 fit using each $\Delta\phi(\ell, E_T^{miss})$ bin as a data point. Similarly to the template fits performed in Sect. 5.3.2, the statistical uncertainties on the bin contents of both the fit target and the templates are included in the denominator of χ^2. Figure 5.18 shows the results of the template fits for both channels. While the contribution from fake leptons is visible in the eγ channel, the fit predicted a negligible fake muon contribution for the $\mu\gamma$ channel.

The scale factors obtained through this method are affected by the normalizations of the components subtracted from data. Therefore, the uncertainties on the fake photon and rare EWK background estimations are directly translated to the uncertainty of the Vγ scale factor. There is also an effect from the data-to-simulation ESF (see Sect. 5.6) applied to the Vγ sample, since the ESF reweights the sample on an event-by-event basis (see Sect. 4.8.3) and thus can change the shapes of the Vγ templates. The effect of these uncertainties on the Vγ scale factor is assessed by repeating the χ^2 minimization 1000 times varying the subtracted distributions and the Vγ template randomly bin-by-by according to the respective uncertainties.

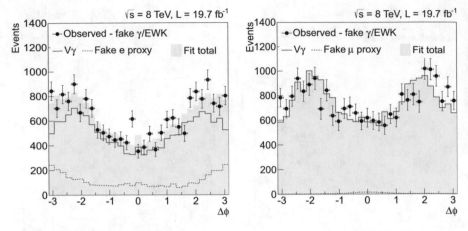

Fig. 5.18 Result of the $\Delta\phi(\ell, E_{\mathrm{T}}^{\mathrm{miss}})$ template fit for the eγ (*left*) and $\mu\gamma$ channel (*right*)

This procedure is performed in the eγ and $\mu\gamma$ channels independently. The resulting scale factors and their uncertainties are

$$a_{\mathrm{V}\gamma(\mathrm{e})} = 1.59 \pm 0.27,$$
$$a_{\mathrm{fake}(\mathrm{e})} = 0.20 \pm 0.07 \tag{5.25}$$

for the eγ channel, and

$$a_{\mathrm{V}\gamma(\mu)} = 1.47 \pm 0.16,$$
$$a_{\mathrm{fake}(\mu)} = 0.01 \pm 0.03 \tag{5.26}$$

for the $\mu\gamma$ channel. The scale factors for the Vγ sample in the two channels agree within uncertainties as expected.

5.5.3   Multiboson and t̄tγ Backgrounds

The WWγ, W Zγ, and t̄tγ processes are considered as rare backgrounds. Dedicated MADGRAPH samples are generated for WWγ and t̄tγ processes. The WZγ events are taken from the inclusive WZ sample generated with PYTHIA. Table 5.1, which is a partial repeat of Table 4.6, lists all the MC simulation samples used for the rare EWK background estimation along with the cross sections used to normalize the samples. The inclusive WW and t̄t samples are included to supplement the WWγ and t̄tγ samples with photon FSR events. The motivation for this treatment is the following. The W boson and the top quark in the WWγ and t̄tγ data sets are generated on-shell without the simulation of their decays at the matrix-element

Table 5.1 List of MC
simulation data sets used for
the rare EWK background
estimation

Name		Generator	σ (pb)	$\mathcal{L}_{\text{eff}}$ (fb^{-1})
$\text{t}\bar{\text{t}}\gamma$		MADGRAPH 5	2.888	596
$\text{WW}\gamma$		MADGRAPH 5	0.528	576.3
$\text{t}\bar{\text{t}}$	Semi-leptonic	MADGRAPH 5	108.7	229.6
	Full-leptonic		26.8	465.6
WW		PYTHIA 6.4	56	178.6
WZ		PYTHIA 6.4	33.21	301.1

See Table 4.6 for details

level, implying that the photons that MADGRAPH generates for these data sets are
always due to ISR. Therefore, events where only the FSR is the source of the prompt
photon must be taken from the inclusive samples and added into the predictions for
completeness.

5.6 Estimations of the Efficiency Scale Factors

5.6.1 Factorization of the Full Selection Efficiency

As discussed in Sect. 4.8.3, when comparing simulation-based background estimations to data, the simulation samples must be corrected for the difference in
the trigger and object selection efficiencies between MC simulation and data. In
Eq. (4.5), the weight given to each simulation event is expressed in terms of the full
selection efficiency ϵ_p^X, where $X = $ data or MC and p is either photon, electron,
or muon. To be fully selected, each object must fire the trigger, be reconstructed
offline, and pass the object selection criteria described in Sects. 4.3, 4.4, and 4.5.
It is therefore convenient to express the full selection efficiency as a product of
conditional probabilities.

Such a factorization of the full efficiency should in principle follow the chrono-
logical order

$$\epsilon^{\text{chron.}} = P(\text{sel.}|\text{reco.}) \cdot P(\text{reco.}|\text{trig.}) \cdot P(\text{trig.}|\text{kin.}), \quad (5.27)$$

where sel., reco., trig., and kin. stand for the offline selection, reconstruction, trigger,
and the kinematic (p_T and η) requirements, respectively. The conditional probability
$P(Y|X)$ describes the rate for a particle passing the requirement X to also pass Y. In
practice, a different factorization is used:

$$\epsilon = P(\text{trig.}|\text{sel.}) \cdot P(\text{sel.}|\text{reco.}) \cdot P(\text{reco.}|\text{kin.}). \quad (5.28)$$

The factorization in Eq. (5.27) is not practical because the probability of a particle to
fire a trigger, $P(\text{trig.}|\text{kin.})$, is a nontrivial quantity that cannot be measured in data.
On the other hand, the probability of a particle to be reconstructed, $P(\text{reco.}|\text{kin.})$,

while also being an unmeasurable quantity, is very close to unity, making its uncertainty negligible. Additionally, the CMS simulation is fairly reliable for a relatively simple task of particle reconstruction, especially for high-p_T objects such as the photons and leptons used in this analysis. In fact, the data over simulation ratio of $P(\text{reco.}|\text{kin.})$ is taken to be unity, since any discrepancy would be fully absorbed in the uncertainty of the other factors. Thus, the measured quantities are the offline selection ESF

$$R_p^{\text{sel}} = \frac{P_p^{\text{data}}(\text{sel.}|\text{reco.})}{P_p^{\text{MC}}(\text{sel.}|\text{reco.})} \tag{5.29}$$

and the trigger ESF

$$R_p^{\text{trig}} = \frac{P_p^{\text{data}}(\text{trig.}|\text{sel.})}{P_p^{\text{MC}}(\text{trig.}|\text{sel.})}. \tag{5.30}$$

The discussion above ignored the fact that multiple objects are involved in each event. For example, the use of the diphoton and the muon–photon triggers implies that $P(\text{trig.}|\text{sel.})$ actually cannot be defined as the probability to fire the trigger, since the event acceptance decision in these triggers depends on more than just the object in consideration. Instead, it should be thought of as the probability to have a trigger object, described in Sect. 5.6.3, passing certain filters.

Another complication that was not discussed is a possible correlation between the selection efficiencies. It is possible that an event in which one object can be fully identified has some properties, such as low overall hadronic activity, that force the other object to also have a higher probability of passing the selection. If such a correlation is significant, the factorized description of the event weight in Eq. (4.5) itself breaks down. Fortunately, studies on simulation show that such a correlation is negligible for the objects used in this analysis.

5.6.2 Offline Selection Efficiency

5.6.2.1 Lepton Selection ESF

Since the object selection criteria are largely standardized in CMS, there exist pre-measured ESFs for photons, electrons, and muons. The electron and muon selection ESFs are determined using the TP technique (see Sect. 5.2) where the tag is a tightly selected electron or muon, and the probe is another electron or muon on which no selection is applied. The number of probes passing and failing the offline selection criteria is counted by integrating the distribution function that is fit to the dilepton invariant mass distribution. The selection efficiency in simulation is calculated directly from the MC truth information. Table 5.2 shows the CMS standard offline selection ESFs for electrons and muons, plotted against p_T and η of the objects.

Table 5.2 Offline selection ESF of electrons and muons

Electron				
p_T range (GeV)	[25, 30]	[30, 40]	[40, 50]	[50, ∞]
$|\eta| < 0.8$	0.986 ± 0.014	1.002 ± 0.003	1.005 ± 0.002	1.004 ± 0.004
$0.8 < |\eta| < 1.4442$	0.959 ± 0.014	0.980 ± 0.003	0.988 ± 0.002	0.988 ± 0.005
$1.556 < |\eta| < 2$	0.941 ± 0.023	0.967 ± 0.007	0.992 ± 0.004	1.000 ± 0.006
$2 < |\eta| < 2.5$	1.020 ± 0.022	1.021 ± 0.007	1.019 ± 0.004	1.022 ± 0.007

Muon				
p_T range (GeV)	[25, 30]	[30, 35]	[35, 40]	[40, 50]
$|\eta| < 0.9$	0.989 ± 0.001	0.987 ± 0.001	0.987 ± 0.001	0.987 ± 0.001
$0.9 < |\eta| < 1.2$	0.995 ± 0.002	0.993 ± 0.002	0.991 ± 0.001	0.991 ± 0.001
$1.2 < |\eta| < 2.1$	1.001 ± 0.001	1.001 ± 0.001	0.998 ± 0.001	0.997 ± 0.001
$2.1 < |\eta| < 2.4$	1.093 ± 0.003	1.075 ± 0.003	1.053 ± 0.002	1.033 ± 0.001

Muon				
p_T range (GeV)	[50, 60]	[60, 90]	[90, 140]	[140, ∞]
$|\eta| < 0.9$	0.988 ± 0.001	0.988 ± 0.001	1.004 ± 0.003	1.017 ± 0.018
$0.9 < |\eta| < 1.2$	0.993 ± 0.001	0.990 ± 0.002	1.010 ± 0.007	1.013 ± 0.035
$1.2 < |\eta| < 2.1$	0.996 ± 0.001	0.993 ± 0.002	1.022 ± 0.006	0.971 ± 0.030
$2.1 < |\eta| < 2.4$	1.019 ± 0.003	1.004 ± 0.005	1.069 ± 0.017	0.900 ± 0.163

5.6.2.2 Photon Selection ESF

The standard ESF for photon selection is further factorized into the ESF for the shower shape and isolation requirements as well as for the electron veto. The first factor can be measured by the electron TP technique, since the distributions in the shower shape and isolation variables of electrons closely resemble those of photons. On the other hand, the second factor must be measured on a sample of real photons. The FSR process $Z \to \mu\mu\gamma$, introduced in Sect. 5.3.2, is also utilized here. It is actually possible to measure the full ESF using the FSR photons, but the rate of $Z \to \mu\mu\gamma$ is orders of magnitude lower than $Z \to ee$, prompting the factorized approach.

The photon selection criteria in Sect. 4.3 differ from the CMS standard selection only in the electron veto. Therefore, the standard value, which is uniformly 0.99 ± 0.01 for barrel photons with $E_T^\gamma > 40\,\text{GeV}$, is used for the first part of the photon selection ESF. For the second part, an original measurement based on a $Z \to \mu\mu\gamma$ TP method is performed as described below.

The SingleMu data set is used for the measurement of the electron veto efficiency. Events with at least two muons with $p_T > 10\,\text{GeV}$ and a barrel photon with $E_T > 40\,\text{GeV}$ are selected. The upper cut on I_{rel} in the muon selection is loosened to 0.4. The photons are required to pass all the selection requirements except for the pixel, GSF, and PF charged hadron vetoes. Since the single-muon trigger in the SingleMu data set has a p_T threshold of 24 GeV, in most of the events, one of the muons has p_T much greater than 10 GeV.

Each $\mu\mu\gamma$ combination that satisfies $m_{\mu\mu} + m_{\mu\mu\gamma} < 180\,\text{GeV}$, where $m_{\mu\mu}$ and $m_{\mu\mu\gamma}$ are the invariant masses of the dimuon and $\mu\mu\gamma$ systems, is then treated independently as one Z boson decay candidate. The cut on the sum of invariant masses is another way of suppressing the ISR events, which was achieved in Sect. 5.3.2 by two consecutive selection cuts. The sample of all $\mu\mu\gamma$ candidates passing the invariant mass selection constitutes the denominator sample. Out of this sample, the $\mu\mu\gamma$ combinations where the photon passes all the electron vetoes are collected as the numerator sample. The denominator and the numerator samples are further split into two bins of $E_\text{T}{}^\gamma$: $40\,\text{GeV} < E_\text{T}{}^\gamma < 50\,\text{GeV}$ and $E_\text{T}{}^\gamma > 50\,\text{GeV}$.

For each sample, binned extended maximum-likelihood fit is performed in the range $60\,\text{GeV} < m_{\mu\mu\gamma} < 120\,\text{GeV}$. The fit proceeds in a very similar manner to the electron TP fit presented in Sect. 5.2.3. The invariant mass distribution of the $\mu\mu\gamma$ system is described by a combination of signal and background templates. Signal template is taken from the invariant mass of the truth-matched $\mu\mu\gamma$ system in the DY simulation data set, and the background template is formed from data by replacing the photon object in the $\mu\mu\gamma$ combination with a hadronic object defined by inverting the isolation requirement of the photon. Figure 5.19 shows the result of the fits. The number of true photons in each sample is inferred from the integral of the signal distribution in $80\,\text{GeV} < m_{\mu\mu\gamma} < 100\,\text{GeV}$.

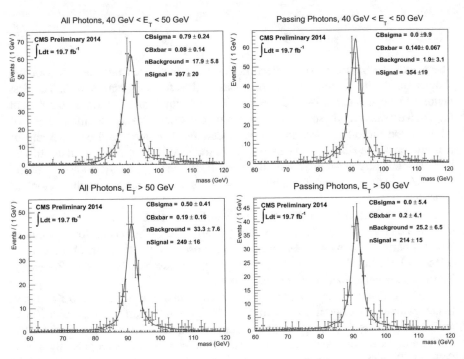

Fig. 5.19 Fits of $m_{\mu\mu\gamma}$ distributions for the electron veto efficiency measurement. Denominator (*left*) and numerator (*right*) fits are shown for $40\,\text{GeV} < E_\text{T}{}^\gamma < 50\,\text{GeV}$ (*top*) and $E_\text{T}{}^\gamma > 50\,\text{GeV}$ (*bottom*)

Table 5.3 Result of photon counting for data and MC simulation

	$40 < E_T^{\gamma} < 50$	$E_T^{\gamma} > 50$
Data efficiency (%)	89.2 ± 1.9	86.0 ± 2.7
Simulation efficiency (%)	88.4 ± 1.5	87.5 ± 1.9
ESF	1.01 ± 0.03	0.86 ± 0.04

Data results come from tag-and-probe fits, whereas the MC results are from direct counting using MC truth information. Errors are statistical

The electron veto efficiency in simulation is calculated from the DY MC simulation mentioned above. The photons in the sample used for constructing the signal template are counted. The calculated efficiencies and the ESF are presented in Table 5.3. Errors on the efficiencies represent 1σ Clopper-Pearson intervals where the efficiencies are treated as binomial proportions. The identical procedure is repeated on the DoubleMu data set as a cross-check. No difference in the final result is observed.

The systematic uncertainty of the measured efficiency is evaluated by repeating the fit procedure with alternative fit models such as the background template described by an analytic function. For each alternative fit, the denominator photon count and the efficiency fraction are compared to the nominal values in Table 5.3, and the differences are taken as uncertainties. In addition, the fits are performed on the DY MC simulation sample to compare the results with the nominal values.

The systematic and statistical uncertainties were combined by a convolution technique employing pseudo-experiments. The denominator value and the efficiency fraction are drawn from normal distributions according to the respective systematic uncertainties, and for each combination the statistical probability distribution for the efficiency is calculated. Such distributions are overlaid for many random draws, resulting in a statistical distribution about the nominal efficiency value smeared by the systematic uncertainties by 1–2%. The 1σ width of this smeared distribution is taken as the total uncertainty on the efficiency.

Aside from the statistical and fit-related uncertainty, there is also a need to assess the effect of pileup, event activity, and the amount of material the photon goes through, since the electron veto based on the number of pixel seeds is sensitive to the number of tracks around the photon, including the conversion tracks. To estimate the possible variation, the ESF is evaluated additionally in bins of photon $|\eta|$, N_{vtx}, and the number of high-p_T jets in the event. An additional 1% uncertainty is assigned to the ESF from this result.

5.6.3   Trigger Efficiency

As mentioned in Sect. 5.6.1, the trigger efficiency is defined as the probability for objects that pass the offline selection criteria to also have a matching HLT object. In practice, this probability is calculated by first identifying an offline object in a sample that is not biased by the trigger under study. Within the set of such offline objects, those that have a matching HLT object are then counted.

From the implementation of the diphoton trigger, the trigger efficiency for the photon in the $e\gamma$ channel, which is required to match the leading leg of the HLT, is

$$P_{e\gamma}^{\gamma}(\text{trig.}|\text{sel.}) = P(\text{Ph22}|\text{Ph36}) \cdot P(\text{Ph36}|\text{EG22}) \cdot P(\text{EG22}|\text{reco.}), \qquad (5.31)$$

where $P(\text{EG22}|\text{reco.})$ is the L1T efficiency, $P(\text{Ph36}|\text{EG22})$ is the efficiency of matching the leading leg object when a L1T match exists, and $P(\text{Ph22}|\text{Ph36})$ is the efficiency of matching the trailing leg object when a leading leg match exists. The last factor is needed since the I_{Trk} requirement is not applied to the leading leg until the last stage of the HLT path. For the electron, the HLT efficiency is simply given by the matching efficiency to the trailing leg object.

The TP method using $Z \rightarrow ee$ events measures the efficiency in the $e\gamma$ channel. The data set on which the measurement is performed is SingleElectron, where one of the two electrons from the Z decay is free of any trigger bias. The denominator of the L1T efficiency is the number of photons in the sample that pass all photon identification criteria except for the electron veto and has a partner tag electron, with the tag-photon two-body invariant mass between 80 and 100 GeV. The electron veto is removed to allow for sufficient statistics to perform this measurement. The effect of this omission is assessed later. The tag is a fully selected electron with $p_T > 40$ GeV that matches the single-electron trigger object. The numerator is the number of photons within the denominator sample that also match the L1T object.

The efficiency of the leading photon leg over the candidate photons is measured in a similar manner, but with two differences on the denominator photons. The first one is the requirement of a matching L1T object, since the leading photon leg is only reconstructed around the L1T e/γ object. The other difference is the requirement of another photon in the event that matches the leading leg object and is separated from the photon under study by $\Delta R > 0.3$. Out of such denominator photons, the ones that have both the leading and the trailing leg objects matched are counted as the numerator. The requirement of another photon in the event is due to the implementation details of the trigger path; the selection on the tracker isolation is applied at the very end of the trigger path that is executed only if there are at least two photon objects.

The SingleElectron data set is also utilized for the electron trigger efficiency measurement. The denominator is the number of candidate-quality electrons which have a partner tag electron with the a two-body invariant mass between 80 and 100 GeV. To guarantee that the trigger path is fully executed, the event is required to have another electron matching the leading leg object and being separated from the electron under study by $\Delta R > 0.3$ is required. The numerator is the number of electrons in the denominator sample that also match the trailing leg object.

The TP methods above are used to obtain pure electron samples in data. In the simulation, the central value for the efficiency is derived by truth-matching the photon and electron objects in the $Z\gamma$ sample. The TP measurements are also performed on the simulation to validate the assumption that the requirement of tag objects in the event does not bias the observed efficiencies. Again, no significant differences are observed.

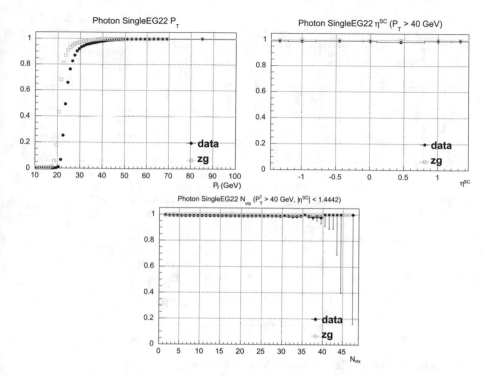

Fig. 5.20 SingleEG22 L1T efficiency over candidate photons

Figures 5.20, 5.21, and 5.22 show the measured trigger efficiencies as functions of p_T and η of the objects and N_{vtx} both for data and simulation. The absolute efficiencies are everywhere lower than unity because the event selection criteria of this analysis is not strictly tighter than the conditions on the trigger. An appreciable difference between the data and simulation efficiencies is observed, both in the overall value and in the dependencies on the variables studied. Since the shape difference is stronger in the η and N_{vtx} dependencies, the data-to-simulation ESF is calculated as a function of these two quantities, while integrating out the p_T dependencies. Note that the p_T thresholds in the object definition, 40 and 25 GeV for the photon and the electron, respectively, are sufficient to place all the candidate objects in the plateau region of the trigger efficiency. Figure 5.23 shows the corresponding ESF for photon and electron objects.

The L1T and leading photon efficiencies are cross-checked by a $Z \to \mu\mu\gamma$ TP measurement, where the full identification criteria, including the electron vetoes, can be used on the photons. The sample used for this measurement is prepared following the prescription for the $Z \to \mu\mu\gamma$ TP study in Sect. 5.6.2. Since it is not common to have two hard photons in dimuon events, it is possible to only measure the leading leg efficiency up to the HCAL isolation, i.e., the numerator is only required to match the leading leg object. Within the possible comparisons, a negligible difference is seen between the efficiency measured by the $Z \to ee$ and $Z \to \mu\mu\gamma$ methods.

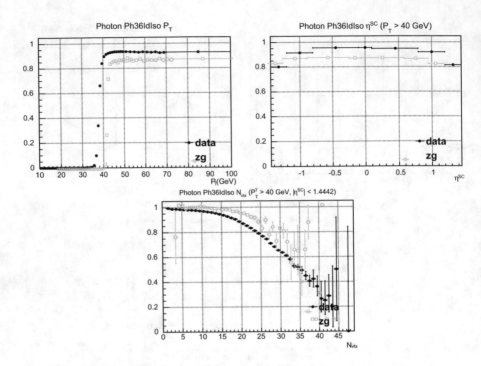

Fig. 5.21 Diphoton HLT leading leg efficiency for candidate photons

The $\mu\gamma$ channel trigger efficiencies are measured using $Z \to \mu\mu\gamma$ and $Z \to \mu\mu$ events exploiting the SingleMu data set. For the seeding L1T bit, which is a muon-e/γ cross trigger, the measurement starts with identifying all the $\mu\mu\gamma$ triplets that pass the criteria used in the $Z \to \mu\mu\gamma$ measurement above. From this sample, the denominator is taken as the number of $\mu\gamma$ pairs with the muon passing the selection criteria in Sect. 4.5. Out of such pairs, those that have both the photon and the muon object matching the L1T e/γ and muon objects are counted in the numerator. The method is validated in simulation by comparing truth-matched result to the TP measurement.

The muon leg of the HLT is reconstructed around the L1T muon object. Thus, for the muon trigger efficiency, the denominator is the number of candidate muons matching the L1T muon object, with a partner muon in the event which also passes the selection criteria and matches the single-muon trigger object. The dimuon invariant mass has to be between 80 and 100 GeV. The numerator is the number of muons in this set that also match the muon leg trigger objects.

The photon leg of the HLT is reconstructed only when at least one muon leg candidate exists, and the reconstruction is again limited to objects around the L1T e/γ candidates. Therefore, the denominator photons are searched for in events where at least one muon-leg object is recorded. Within such events, photons from the

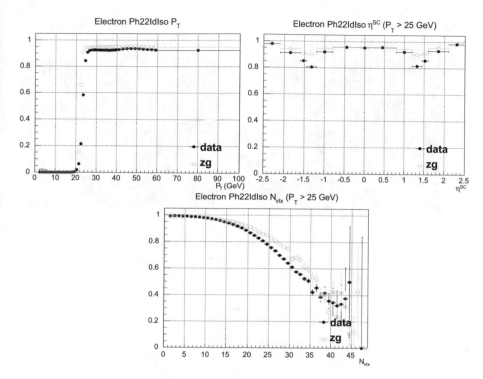

Fig. 5.22 Diphoton HLT trailing leg efficiency for candidate electrons

$Z \rightarrow \mu\mu\gamma$ process are identified, and the fraction of such photons that have matching photon-leg objects are measured.

Figures 5.24, 5.25, and 5.26 show the measured trigger efficiencies as functions of p_T and η of the objects and N_{vtx}. The measurements based on $Z \rightarrow \mu\mu\gamma$ suffer from low statistics. On the other hand, both the L1T and photon leg efficiencies do not show obvious signs of data-to-simulation ratio depending on the variables studied. Therefore, for simplicity, a single scale data-to-simulation ESF, 0.995 ± 0.002(stat.) and 0.995 ± 0.001(stat.), are assigned to the L1T seed and the photon leg, respectively. The scale factor for the muon leg, on the other hand, is derived as a function of muon η, as shown in Fig. 5.23.

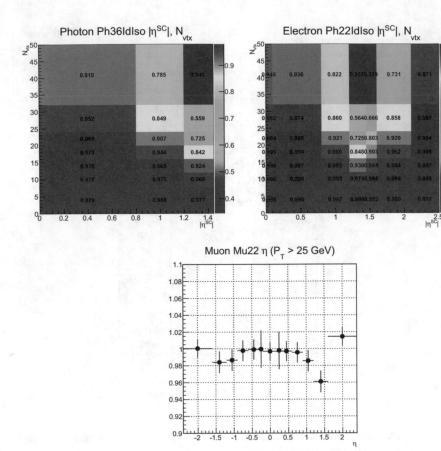

Fig. 5.23 Trigger efficiency data over MC scale factors for photons, electrons, and muons

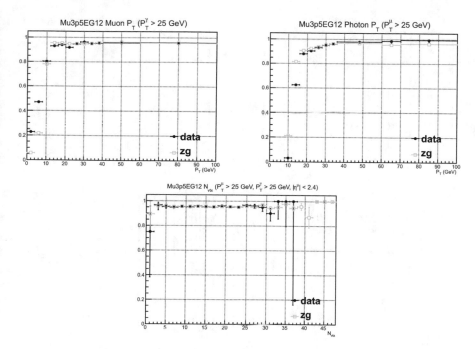

Fig. 5.24 Muon-e/γ L1T efficiency for candidate muons and photons

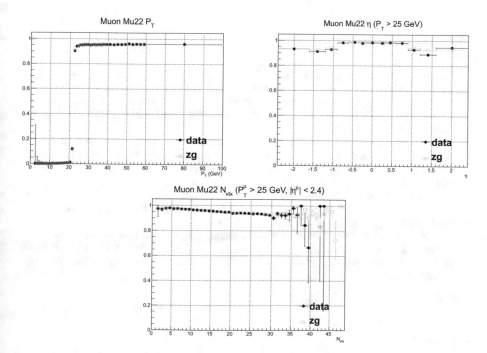

Fig. 5.25 Muon–photon HLT efficiency for candidate muons

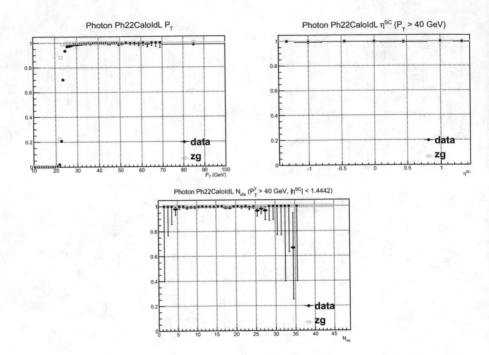

Fig. 5.26 Muon–photon HLT efficiency for candidate photons

References

1. Cranmer, K.S.: Kernel estimation in high-energy physics. Comput. Phys. Commun. **136**, 198–207 (2001). doi:10.1016/S0010-4655(00)00243-5. arXiv: hep-ex/0011057 [hep-ex]
2. Campbell, J.M., Ellis, R.K., Williams, C.: Vector boson pair production at the LHC. J. High Energy Phys. **1107**, 018 (2011). doi:10.1007/JHEP07(2011)018. arXiv: 1105.0020 [hep-ph], FERMILAB-PUB-11-182-T
3. CMS Collaboration. Measurement of the Wγ and Zγ inclusive cross sections in pp collisions at $\sqrt{s}$ = 7 TeV and limits on anomalous triple gauge boson couplings. Phys. Rev. D **89**, 092005 (2014). doi:10.1103/PhysRevD.89.092005. arXiv: 1308.6832 [hep-ex], CMS-EWK-11-009, CERN-PH-EP-2013-108
4. ATLAS Collaboration. Measurements of Wγ and Zγ production in pp collisions at $\sqrt{s}$ = 7 TeV with the ATLAS detector at the LHC. Phys. Rev. D **87**(11), 112003 (2013). doi:10.1103/PhysRevD.87.112003. arXiv: 1302.1283 [hep-ex]
5. Gleisberg, T., et al.: Event generation with sherpa 1.1. J. High Energy Phys. **0902**, 007 (2009). doi:10.1088/1126-6708/2009/02/007. arXiv: 0811.4622 [hep-ph], FERMILAB-PUB-08-477-T, SLAC-PUB-13420, ZU-TH-17-08, DCPT-08-138, IPPP-08-69, EDINBURGH-2008-30, MCNET-08-14

Chapter 6
Results and Interpretations

6.1 Results

Having determined all components of the background, the observed data are compared to the background estimation. Figures 6.1, 6.2, 6.3 show the observed distributions of the E_T^{γ}, lepton p_T (p_T^{ℓ}), E_T^{miss}, M_T, H_T, and the number of jets (N_{jet}) in the $e\gamma$ and $\mu\gamma$ channels, together with the stacked background estimations. Two benchmark signal event distributions, one each from the TChiWg and T5Wg models, are also shown in the figures. The TChiWg point is for $m_{\tilde{\chi}} = 300\,\text{GeV}$, which has the nominal cross section of 0.146 pb, and the T5Wg point is for $m_{\tilde{g}} = 1000\,\text{GeV}$ and $m_{\tilde{\chi}} = 425\,\text{GeV}$, which has the nominal cross section of 0.0122 pb. In the figures and the remainder of this section, the displayed uncertainties are a combination of the systematic and statistical uncertainties, added in quadrature. Details on the determination of the systematic uncertainties are given in Sect. 6.3.

The data and the background estimation are compared in the SUSY signal search region $E_T^{miss} > 120\,\text{GeV}$ and $M_T > 100\,\text{GeV}$. To gain sensitivity for multiple possible SUSY scenarios, the signal region is further divided into two E_T^{γ} bins of below and above 80 GeV, three H_T bins with boundaries at 100 and 400 GeV, and three E_T^{miss} bins with boundaries at 200 and 300 GeV. The segments of the kinematic parameter space divided this way are referred to as signal region counting bins. Figure 6.4a, b shows the M_T distribution for the $e\gamma$ and $\mu\gamma$ channels after requiring $E_T^{miss} > 120\,\text{GeV}$. The panels (c) and (d) in the same figure show the opposite, i.e., the E_T^{miss} distribution with the requirement of $M_T > 100\,\text{GeV}$. The distributions of E_T^{γ} and H_T after applying both $E_T^{miss} > 120\,\text{GeV}$ and $M_T > 100\,\text{GeV}$ requirements are also shown in Fig. 6.4.

Table 6.1 summarizes the SM expectations, the observed number of events, and the expected number of events from the two benchmark signal samples. Two E_T^{γ} bins are combined in the table. Figure 6.5 shows the total background expectation

© Springer International Publishing AG 2017
Y. Iiyama, *Search for Supersymmetry in pp Collisions at* $\sqrt{s} = 8$ TeV *with a Photon, Lepton, and Missing Transverse Energy*, Springer Theses,
DOI 10.1007/978-3-319-58661-8_6

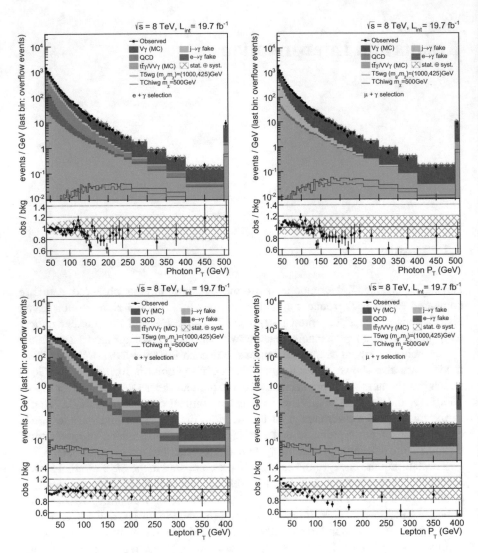

Fig. 6.1 E_T^γ (*top*) and p_T^ℓ (*bottom*) distributions in the eγ channel (*left*) and the $\mu\gamma$ channel (*right*) with stacked background estimations

and the event yield of each signal region counting bin. The data are in good agreement with the background predictions, and no excess of events in the signal region is observed.

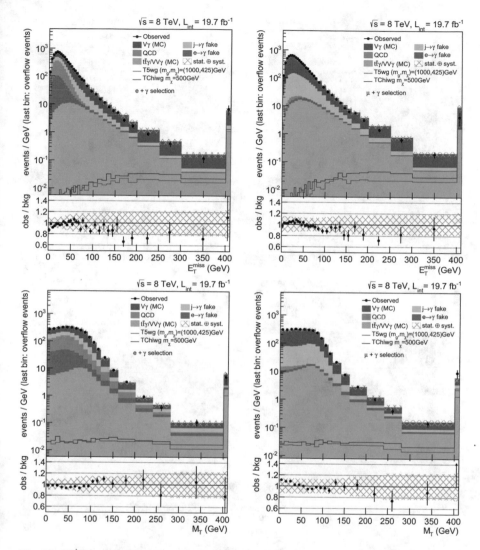

Fig. 6.2 E_T^{miss} (*top*) and M_T (*bottom*) distributions in the eγ channel (*left*) and the $\mu\gamma$ channel (*right*) with stacked background estimations

6.2 Signal Expectations

To understand the implication of the result of this search for the MSSM, the behavior of the signal model events under the event selection criteria and the background estimation methods are studied. As mentioned in Sect. 4.8.1, the simulated and ESF-corrected signal events are subjected to the full event selection applied to data. Additionally, the possibility that a "contamination" of signal events exists in the

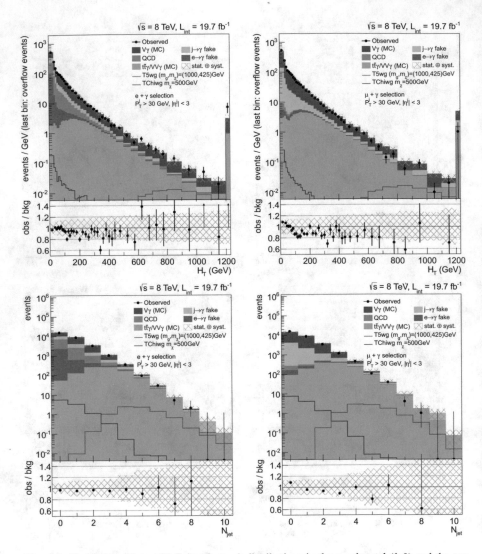

Fig. 6.3 H_T (*top*) and jet multiplicity (*bottom*) distributions in the eγ channel (*left*) and the $\mu\gamma$ channel (*right*) with stacked background estimations

proxy samples used for the fake background estimations must be considered, since the presence of signal events in the control samples can cancel out a deviation of the observed data from the SM prediction. To evaluate the effect of signal contamination, proxy samples are also formed from the signal simulation events, scaled by the appropriate transfer factors, and subtracted from the total event yield. The result is then scaled to the cross section for the respective mass point to obtain the signal expectation.

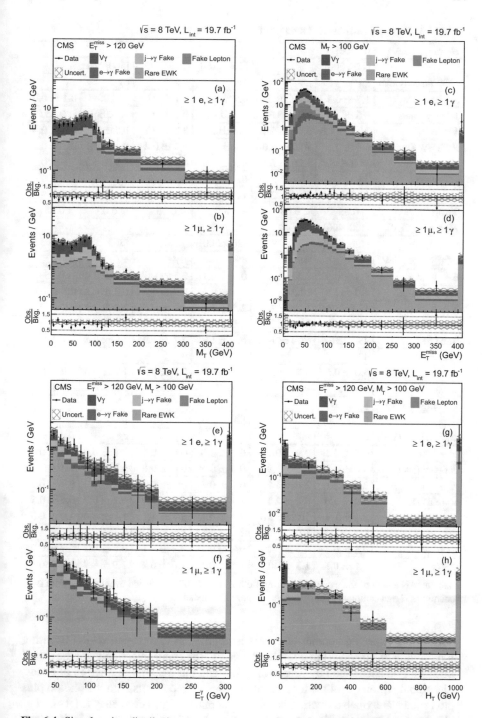

Fig. 6.4 Signal region distributions comparing the observed data to the background estimation. M_T (**a**), (**b**) is shown after applying $E_T^{miss} > 120\,\text{GeV}$ and E_T^{miss} (**c**), (**d**) after $M_T > 100\,\text{GeV}$. Both requirements are applied for the E_T^{γ} (**e**), (**f**) and H_T (**g**), (**h**) distributions

Table 6.1 Expected and observed number of events in bins of H_T and E_T^{miss}

H_T range (GeV)	[0, 100]					
	[120, 200]		[200, 300]		[300, ∞]	
E_T^{miss} range (GeV)	eγ	μγ	eγ	μγ	eγ	μγ
e→γ fakes	1.7 ± 0.3	2.1 ± 0.3	0.1 ± 0.0	0.1 ± 0.0	< 0.05	0.1 ± 0.0
Jet→γ fakes	6.3 ± 1.3	8.8 ± 3.7	< 0.05	0.9 ± 0.7	< 0.05	< 0.05
Vγ	21.5 ± 4.6	33.8 ± 5.0	4.8 ± 1.7	3.8 ± 0.7	0.9 ± 0.3	1.1 ± 0.4
Rare EWK	6.9 ± 2.6	10.9 ± 4.0	1.4 ± 0.6	0.9 ± 0.4	0.1 ± 0.1	0.3 ± 0.2
Total SM	36.3 ± 5.4	55.6 ± 7.4	6.3 ± 1.8	5.8 ± 1.1	1.1 ± 0.3	1.5 ± 0.5
Observed	45	51	6	5	1	1
TChiWg	23.2 ± 2.3	31.7 ± 2.8	15.8 ± 1.6	22.4 ± 2.1	3.4 ± 0.5	3.7 ± 0.6
T5Wg	< 0.05	< 0.05	< 0.05	< 0.05	< 0.05	< 0.05
H_T range (GeV)	[100, 400]					
	[120, 200]		[200, 300]		[300, ∞]	
E_T^{miss} range (GeV)	eγ	μγ	eγ	μγ	eγ	μγ
e→γ fakes	5.6 ± 0.9	7.2 ± 1.1	0.4 ± 0.1	0.7 ± 0.1	0.1 ± 0.0	< 0.05
Jet→γ fakes	4.0 ± 1.0	12.3 ± 5.1	0.5 ± 0.3	0.5 ± 0.6	0.2 ± 0.2	0.5 ± 0.6
Vγ	12.7 ± 2.6	15.2 ± 2.3	2.5 ± 1.2	1.9 ± 0.5	0.8 ± 0.3	0.5 ± 0.2
Rare EWK	21.0 ± 7.9	34.0 ± 12.9	2.9 ± 1.1	4.7 ± 1.7	0.6 ± 0.3	0.7 ± 0.3
Total SM	43.2 ± 8.4	68.8 ± 14.1	6.3 ± 1.7	7.7 ± 1.9	1.7 ± 0.5	1.7 ± 0.7
Observed	42	71	5	9	1	4
TChiWg	3.0 ± 0.5	3.7 ± 0.5	3.0 ± 0.5	3.7 ± 0.8	1.2 ± 0.3	1.9 ± 0.4
T5Wg	< 0.05	< 0.05	< 0.05	< 0.05	0.1 ± 0.0	0.1 ± 0.0
H_T range (GeV)	[400, ∞]					
	[120, 200]		[200, 300]		[300, ∞]	
E_T^{miss} range (GeV)	eγ	μγ	eγ	μγ	eγ	μγ
e→γ fakes	0.7 ± 0.1	0.9 ± 0.1	0.2 ± 0.0	0.4 ± 0.1	0.1 ± 0.0	0.1 ± 0.0
Jet→γ fakes	0.6 ± 0.4	1.1 ± 0.9	0.3 ± 0.3	0.6 ± 0.6	0.3 ± 0.3	< 0.05
Vγ	2.2 ± 0.6	2.6 ± 1.2	0.6 ± 0.3	0.6 ± 0.3	0.4 ± 0.2	0.3 ± 0.2
Rare EWK	4.2 ± 1.6	5.7 ± 2.1	1.1 ± 0.5	2.1 ± 0.9	0.3 ± 0.2	0.5 ± 0.3
Total SM	7.7 ± 1.7	10.2 ± 2.6	2.1 ± 0.6	3.7 ± 1.1	1.1 ± 0.4	1.0 ± 0.3
Observed	8	10	1	1	1	0
TChiWg	0.3 ± 0.2	0.3 ± 0.2	0.2 ± 0.1	0.2 ± 0.1	0.7 ± 0.2	0.7 ± 0.2
T5Wg	1.1 ± 0.1	1.3 ± 0.1	1.5 ± 0.1	1.6 ± 0.2	2.9 ± 0.4	3.5 ± 0.4

Each entry is a sum of the event yields from both E_T^γ bins. The expected event yields from two benchmark signal samples described in the text are also listed

While the expected number of events is counted in each signal region counting bin individually, an overall trend of how efficient the event selection is on the signal models can be understood by calculating for each mass point the fraction of events in the signal region out of the total number of generated events with the photon plus lepton final state. Such a fraction is called acceptance times efficiency ($\mathcal{A} \times \epsilon$).

Figure 6.6 shows the $\mathcal{A} \times \epsilon$ value for each point of the GMSB model in the plane of gluino–wino mass plane. Since the number of generated events for the strong

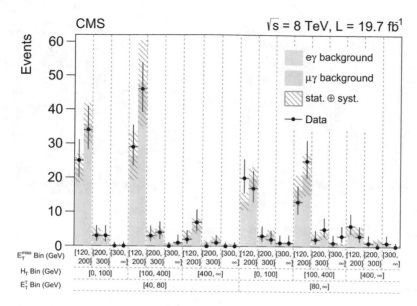

Fig. 6.5 Background expectations and event yields of all signal region counting bins

and electroweak production processes was not proportional to their respective cross sections, $\mathcal{A} \times \epsilon$ for each point of this model was defined using a formula similar to Eq. (4.4):

$$\mathcal{A} \times \epsilon = \frac{\left(\sum_P \sigma_P n_P\right)^2}{\sum_P \sigma_P^2 n_P} \bigg/ \frac{\left(\sum_P \sigma_P N_P^{\gamma\ell}\right)^2}{\sum_P \sigma_P^2 N_P^{\gamma\ell}} \tag{6.1}$$

where P, σ_P, and $N_P^{\gamma\ell}$ are identical to Eq. (4.4) and n_P is the number of events from the process P in the signal region. Figures 6.7 and 6.8 show the $\mathcal{A} \times \epsilon$ values for the TChiWg and T5Wg points.

6.3 Systematic Uncertainties

The systematic uncertainties in the SM background estimations and the signal expectation are identified and evaluated. Table 6.2 summarizes the sources of systematic uncertainties considered, which are described below. The third and fourth columns of the table show the relative magnitude of the uncertainty with respect to the expected event yield in the signal region. For the uncertainties in background estimation, the third column gives the uncertainty relative to the estimation of the corresponding background component, while the fourth column is that relative to the total background estimation. For the uncertainties in signal expectation, the third

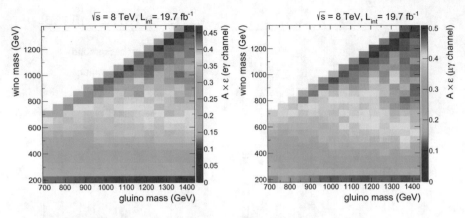

Fig. 6.6 GMSB acceptance times efficiency in the gluino–wino mass plane

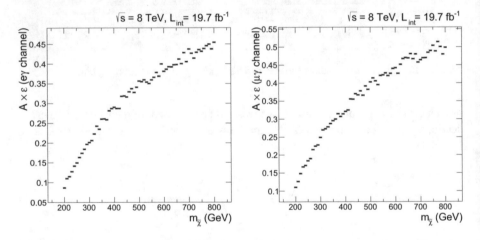

Fig. 6.7 TChiWg acceptance times efficiency

column presents the relative uncertainty with respect to the total signal expectation calculated independently for each mass point of the SUSY signal data sets. In case these relative uncertainties differ significantly from one sample to another, which can be caused by small numbers of expected events in some samples used to evaluate the systematic uncertainties, the range from the minimum to the maximum relative uncertainty is shown. The descriptions of the items in the table are the following.

Vγ Scale Factor   One of the two biggest sources of uncertainties is the scale factor $a_{\mathrm{V}\gamma}$ in Eqs. (5.25) and (5.26) for the Vγ background. Normalization errors of the fake background estimations are absorbed in the uncertainty of the Vγ scale factor through the error estimation process described in Sect. 5.5.2. In other words, while the fake backgrounds have only minor direct contributions to the signal region, their uncertainties in the low-$E_{\mathrm{T}}^{\mathrm{miss}}$ control region are propagated to the signal region through the Vγ estimation.

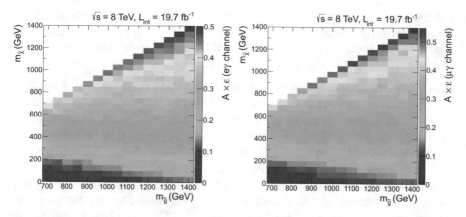

Fig. 6.8 T5Wg acceptance times efficiency

Table 6.2 Summary of the systematic uncertainties considered in the final interpretation of the analysis result

| Name | Sample | Rel. uncertainty (%) | |
		Sample	Total
Rare background rate	Rare EWK	50	19
Vγ scale factor	Vγ	14	6
Proxy sample shape	Fake	20–27	5
Trigger and ID efficiency	Rare EWK	8	3
JES	EWK	0–6	2
Vγ shape	Vγ	5	2
Integrated luminosity	Rare EWK	2.6	1
JER	EWK	0–2	1
JES	Signal	0–22	–
JER	Signal	0–17	–
Trigger and ID efficiency	Signal	8	–
Initial-state radiation	Signal	0–5	–
Integrated luminosity	Signal	2.6	–
Renormalization scale and PDF	Signal	4–41	–

For each uncertainty in background estimations, its magnitude relative to the expected contribution from affected background component in the signal region is shown in the third column. A range is given in case there are multiple affected components whose relative uncertainty values differ significantly. The fourth column shows the relative uncertainty with respect to the total background expectation in the signal region. For each uncertainty in signal expectations, its magnitude relative to the expected event yield in the signal region is shown in the third column. A range is given for the sources of uncertainty that affect various mass points of the SUSY signal data sets differently

Rare Background Rate The other biggest source of uncertainty is the theoretical cross sections used to normalize the rare EWK background simulations. A 50% uncertainty is uniformly assigned to the theoretical cross sections of the rare EWK background processes used to normalize the MC simulation samples. The process that has the largest contribution in the signal region in this category is $t\bar{t}\gamma$. Since the relative difference of the observed cross section of this process, measured by CMS [1], to calculation is 30%,[1] the value of 50% is a conservative estimate of the uncertainty.

Proxy Sample Shape The uncertainties in the minor contributions of the fake backgrounds to the signal region are also evaluated. Since the overall scale uncertainties are already accounted for, only the effect of a possible mis-modeling of the fake background distribution shapes by the proxy samples has to be evaluated. For the electron-to-photon fakes, a flat 20% error derived in Sect. 5.2.9 is assigned to the event yield estimations of this background in all signal region bins. For the jet-to-photon and jet-to-lepton fakes, the shape uncertainties are assessed by significantly loosening the proxy object definitions to contain even more hadronic contributions. The fractions of events in the signal region $E_{\mathrm{T}}^{\mathrm{miss}} > 120\,\mathrm{GeV}$ and $M_{\mathrm{T}} > 100\,\mathrm{GeV}$ of these new proxy samples are then compared to the corresponding values from the nominal proxy samples, and the relative differences of these fractions are used as the estimations of the error.

Trigger and ID Efficiency The ESF affects the normalization of the rare EWK background and the signal expectations. However, for both samples, the uncertainty in the cross section dominates any possible errors on the ESF. The uncertainty in the signal cross section is discussed below. A uniform 8% uncertainty is assigned to both samples.

Jet Energy Scale and Jet Energy Resolution For MC simulation-based background estimations and the SUSY signal simulations, a potential difference between the simulation and the data in the jet energy scale (JES) and jet energy resolution (JER) must be considered as systematic uncertainties. JES and JER affect not only the variables directly related to jets, such as H_{T} and N_{jet}, but also $E_{\mathrm{T}}^{\mathrm{miss}}$ and therefore M_{T}, through the $E_{\mathrm{T}}^{\mathrm{miss}}$ correction using the jet momenta described in Sect. 3.4.5. Changes in the H_{T} and $E_{\mathrm{T}}^{\mathrm{miss}}$ values in particular induce the so-called event migrations, where the total number of expected events in the signal region remains unchanged but the events get redistributed among the signal region counting bins. To assess the effect of the JES uncertainty, the $E_{\mathrm{T}}^{\mathrm{miss}}$, M_{T}, and H_{T} values are recalculated for each event in the $\mathrm{V}\gamma$, rare EWK, and signal samples with the JES uniformly scaled by $\pm 1\sigma$. The shift in the expected event yield in each counting bin is taken as the estimate of the systematic uncertainty. For the JER uncertainty evaluation, the p_{T} of each jet is randomly smeared by a Gaussian whose width is dependent on the amount of pileup as well as p_{T} and $|\eta|$ of the jet. A CMS-standard formula and its parameters are used to determine the width from these variables. The result of this smearing on the MC simulation samples is then propagated again through recalculation of $E_{\mathrm{T}}^{\mathrm{miss}}$, M_{T}, and H_{T}. In some statistically limited counting bins, the effect of event migration can be at tens of percent, even though its actual contribution to the overall uncertainty is small.

---

[1] The two values agree within errors.

VγShape The uncertainty in the shape of the distributions is also relevant for the Vγ background. The most important uncertainty would be in the E_T^{miss} distribution shape of the Wγ sample. However, since E_T^{miss} is sensitive to signal, it is not possible to directly compare the E_T^{miss} distributions in MC and data to evaluate the goodness of the modeling. Therefore, the muon p_T spectrum of the Zγ MADGRAPH sample is instead compared to that of $\mu\mu\gamma$ events in data, under the assumption that the similarity of the Wγ and Zγ processes will lead to a similar mis-modeling, if any exists.

For data, $\mu\gamma$ channel candidate events with $E_T^{miss} < 70$ GeV are further selected by a requirement for an additional muon passing a loose selection. To enhance the dimuon purity, only events with a dimuon invariant mass between 80 GeV and 100 GeV are employed. From the muon p_T distribution in this sample, the estimated contributions from dimuon events in the fake photon and the rare EWK backgrounds are subtracted. The resulting histogram is then compared to the p_T distribution from the Zγ sample selected with the identical conditions. The Zγ distribution is scaled by the scale factor from Eq. (5.26).

The two distributions are statistically consistent, as shown on the left-hand side of Fig. 6.9. Nevertheless, to propagate the small discrepancy back to the Wγ shape uncertainty, the Vγ background is reweighted with the leading lepton p_T by the ratio of the two distributions, given on the right-hand side of Fig. 6.9. The difference between the modified and nominal background prediction is then used as the systematic uncertainty.

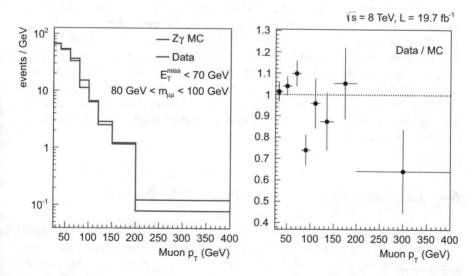

Fig. 6.9 *Left*: Muon p_T distributions in $\mu\mu\gamma$ events in data and Zγ samples. The estimated contribution from fake photon and rare EWK events is subtracted from data, and the Zγ sample is scaled by the factor given in Eq. (5.26). *Right*: Ratio of the two distributions

Integrated Luminosity The integrated luminosity, which is used to normalize the rare EWK background and the signal, is given a 2.6% uncertainty following the CMS-wide recommendation [2].

Initial-State Radiation The p_T spectrum of the ISR simulated by MADGRAPH is known to have a somewhat longer tail than what is observed. Since a very high-p_T ISR can boost the remainder of the hard scattering, affecting quantities such as E_T^{miss}, the effect of the ISR is evaluated on the signal samples. The TChiWg and T5Wg data sets are MADGRAPH-based, which allows the use of an established prescription in [3]. In this method, the uncertainty is evaluated by reweighting the signal events by the transverse momentum of the initial di-sparticle system (p_T^{SUSY}). The following weights are used: unity for $p_T^{SUSY} \leq 120\,\text{GeV}$; 0.95 for $120\,\text{GeV} < p_T^{SUSY} \leq 150\,\text{GeV}$; 0.9 for $150\,\text{GeV} < p_T^{SUSY} \leq 250\,\text{GeV}$; 0.8 for $p_T^{SUSY} > 250\,\text{GeV}$. The resulting expected event yields are then compared to the nominal expectations, and the difference is used as the uncertainty for each counting bin. For the GMSB model, which is generated exclusively with PYTHIA, a flat 5% uncertainty is assigned to all counting bins.

Renormalization/Factorization Scale and PDF The dominant uncertainty for the expected signal yields is in the renormalization scale and PDF used in the cross section. For the models TChiWg and T5Wg, the magnitude of the uncertainties is calculated together with the cross section central values by the LHC SUSY cross section working group [4]. For the GMSB model, the uncertainty in the cross section is calculated as $\sqrt{\sigma_{PDF}^2 + \sigma_{\mu_F}^2}$, where σ_{PDF} is the uncertainty in the PDF, and σ_{μ_F} the one on the factorization scale. The PDF uncertainty is evaluated by recalculating the cross section with alternative PDF sets, where the parameters of the PDF fit are shifted by $\pm 1\sigma$ along the eigenvector directions of the Hessian matrix. The μ_F uncertainty was evaluated by repeating the calculation at factorization scales $\frac{1}{2}m_Z$ and $2m_Z$ and taking the difference from the nominal cross section. A large uncertainty is seen in very high-mass points where this search is not sensitive to. Since this uncertainty is on the theoretical prediction itself rather than on the experimental methods, it is treated separately in the interpretation of the results described in Sect. 6.5.

6.4   CL$_s$ Limit with One-Sided Profile Likelihood

Since no excess of events beyond the SM expectation is observed in any of the signal region bins, the results from this search are interpreted in terms of cross section upper limits on the signal models. The "LHC-style" CL$_s$ limits [5] which use the one-sided profiled likelihood as the test statistic are calculated. The following is a general description of this method.

CL$_s$ [6, 7] is a measure of confidence for excluding a signal hypothesis in view of some observation. The observation in the current case is the event yields in the signal region bins. The confidence level is calculated using a test statistic q whose value is larger for more signal-like data. Denoting the value of q for the result of the experiment as q^{obs}, CL$_s$ is given as the probability of observing a higher q value than q^{obs} under the signal hypothesis, normalized by the same probability under the null (background-only) hypothesis.

Usually, instead of quoting the CL$_s$ value for one specific signal hypothesis, a continuous family of hypotheses where the signal strength is scaled by a parameter μ is considered. In other words, models where the mean expectation for the observed number of events is symbolically $b + \mu s$, where b is the background expectation and s is the nominal signal expectation, are tested against the null hypothesis, corresponding to $\mu = 0$. Often μ itself is called the signal strength, and the hypothesis with $\mu = r$ is denoted H_r. If for $\mu = r_{95}$

$$\mathrm{CL}_s(q^{obs}; r_{95}) = \frac{\mathrm{Prob}(q \geq q^{obs}; H_{r_{95}})}{\mathrm{Prob}(q \geq q^{obs}; H_0)} = 0.05, \tag{6.2}$$

r_{95} is called the 95% upper limit for the signal strength, and H_r with $r > r_{95}$ is considered excluded at the 95% confidence level. Figure 6.10 shows an example of the calculated CL$_s$ values for different μ values at a TChiwg point with $m_{\tilde{\chi}} = 500\,\mathrm{GeV}$. The 95% upper limit is obtained by approximating CL$_s(q^{obs}; \mu)$ locally by a falling exponential and calculating the point where the curve crosses the horizontal line CL$_s = 0.05$.

The term probability in the above discussion must be clarified. First, note that the test statistic q can depend on μ for its computation. As a function of the input data, it should hence be denoted more properly as q_μ(data). The data here can actually be the observed event yields, or a possible outcome of the hypothetical experiment

Fig. 6.10 Distribution of CL$_s$ values for different signal strength μ at a TChiwg point with $m_{\tilde{\chi}} = 500\,\mathrm{GeV}$. The function CL$_s(q^{obs}; \mu)$ is locally approximated by an exponential fit function (*red curve*) to obtain the 95% confidence level upper limit r^{95}, which is indicated by an *arrow* pointing to the horizontal axis

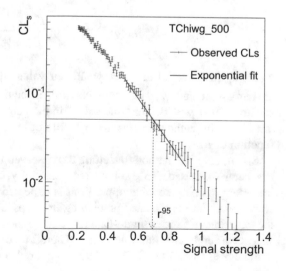

under H_μ. In the latter case, the yield n_i in the signal region bin i follows the Poisson distribution

$$\text{Pois}(n_i; b_i + \mu s_i) = \frac{(b_i + \mu s_i)^{n_i}}{n_i!} e^{-b_i - \mu s_i} \tag{6.3}$$

where b_i and s_i are the background and nominal signal expectations in bin i. Thus, the probability for observing the yields $\{n_i\}$ is well defined, and is given by the product of the right-hand side of Eq. (6.3) for all bins. To be precise, systematic uncertainties in determining $\{b_i\}$ and $\{s_i\}$ must be accounted for. The expected yields are therefore actually functions of nuisance parameters θ that encode the uncertainties, and the product of Poisson probabilities must be multiplied by the probability for θ to take a specific value. In short, the function

$$\mathcal{L}(\{n_i\}|\mu, \theta) = \prod_i \text{Pois}(n_i; b_i(\theta) + \mu s_i(\theta)) p(\theta) \tag{6.4}$$

gives the likelihood of $\{n_i\}$ for a given μ and θ. The distribution of $\{n_i\}$ can then be translated to the distribution of q_μ, from which the probabilities used in the CL_s calculation are inferred.

Due to the existence of the nuisance parameters, neither the distributions of $\{n_i\}$ nor q_μ can be found analytically. Therefore, in a practical calculation, a large number of $\{n_i\}$ sets are generated with Monte Carlo technique for multiple H_μ. Such randomly generated simulated outcomes of the experiment are called pseudo-data. The numerator and denominator probabilities for the CL_s calculations can be obtained from simple integrals of the histograms of q_μ filled with pseudo-data.

The test statistic q_μ that is used below is the one-sided profile likelihood ratio [5]:

$$q_\mu(\{n_i\}) = -2 \ln \frac{\mathcal{L}(\{n_i\}|\mu, \hat{\theta}_\mu)}{\mathcal{L}(\{n_i\}|\hat{\mu}, \hat{\theta})}, \tag{6.5}$$

where $\hat{\theta}_\mu$ is the set of nuisance parameter values that maximizes the likelihood in the numerator, and $(\hat{\mu}, \hat{\theta})$ globally maximizes the likelihood for given $\{n_i\}$ under the constraint $0 < \hat{\mu} \le \mu$. The "one-sided" in the name of the test statistic refers to this last constraint, which ensures that q_μ will not have a minimum when regarded as a function of μ.

So far, only the method for setting the observed upper limit of the signal strength has been discussed. To evaluate the sensitivity of an experiment, it is equally important to compute what upper limit is expected under H_0. For this purpose, the above procedure is repeated pretending that a pseudo-data from H_0 is the result of the experiment. Since pseudo-data has a finite distribution, the 95% upper limit r_{95} will accordingly take multiple values. Therefore, expected limits are usually quoted at the 2.5%, 16%, 50%, 84%, and 97.5% quantiles of the r_{95} distribution.

The calculation of the expected limits can be extremely computation-intensive. In the most straightforward implementation, if M pseudo-data points from H_0 are used as observations and N_j computations are needed for the signal strengths $\{\mu_j\}$ to accurately calculate $\mathrm{CL_s}(q_{\mu_j}^{\mathrm{pseudo\text{-}obs}}; \mu_j)$, the required number of times the test statistic is evaluated scales like $\sim M \times \sum_j N_j$. Even with modern computers, such calculations can easily take days to weeks to finish. Therefore a much more efficient algorithm is implemented in the standard CMS limit-setting tool, which was also used to obtain the results in the next sections.

The algorithm used in CMS starts with the generation of q_{μ_i} distributions for H_{μ_i} and H_0. Both distributions will be used in $\mathrm{CL_s}$ calculation later, but the latter distribution $f(q_{\mu_i}; H_0)$ can in fact be also regarded as the distribution of possible observations $q_{\mu_i}^{\mathrm{pseudo\text{-}obs}}$ under H_0. Then, since $\mathrm{CL_s}(q_\mu^{\mathrm{obs}}; \mu)$ is a monotonically decreasing function of q_μ^{obs} for fixed μ, $\mathrm{CL_s}(q_{\mu_i}^k; \mu_i)$ where $q_{\mu_i}^k$ is the k-quantile of $f(q_{\mu_i}; H_0)$ coincides with the k-quantile value of the $\mathrm{CL_s}$ distribution under H_0. As $\mathrm{CL_s}(q_\mu^{\mathrm{obs}}; \mu)$ is also a monotonically decreasing function of μ for fixed $\{n_i\}$, if the k-quantile of $\mathrm{CL_s}$ is at 0.05 for a signal strength r_{95}^k, then a fraction k of observations made under H_0 will have their 95% upper limit below r_{95}^k. Therefore, the expected limits are given by r_{95}^k for $k = 0.025, 0.16, 0.5, 0.84, 0.975$.

The software package used for the calculation of the profile likelihood ratio was RooStats [8] version 5.34.04-cms2. Since the sources of the systematic uncertainties described in Sect. 6.3 are all multiplicative factors to the expected event yields, they were all treated as nuisance parameters with a log-normal probability distribution with the peak at the nominal value and width given by the corresponding systematic uncertainty. The statistical uncertainty of each background estimation, which also becomes a systematic uncertainty in the context of Eq. (6.3), was modeled by a gamma distribution

$$\frac{\beta^\alpha}{\Gamma(\alpha)} x^{\alpha-1} e^{-\beta x}, \tag{6.6}$$

where x is the expected event yield, α is the original number of events in the control sample, and β is the nominal expectation divided by α, i.e., the averaged transfer factor.

6.5 Interpretations

The $\mathrm{CL_s}$ calculation is performed by constructing the likelihood in Eq. (6.4) from the observed event yields and background estimations in the signal region counting bins. Figures 6.11 and 6.12 show the computed 95% $\mathrm{CL_s}$ cross section upper limits on the GMSB, TChiwg, and T5Wg models. The black and red curves shown in Fig. 6.11 and on the right-hand side of Fig. 6.12 are where the observed and 50%-quantile expected upper limits are equal to the calculated cross section given in

Fig. 6.11 95% CL$_s$ cross
section upper limits and
exclusion contours for the
GMSB model

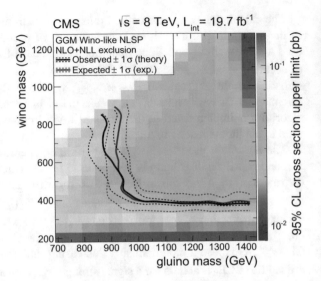

Sect. 4.8.1. Such curves are called 95% confidence level exclusion contours. The
TChiWg model was scanned in one dimension of the degenerate mass of $\widetilde{\chi}^\pm$ and $\widetilde{\chi}^0$.
In this one-dimensional scan, shown on the left-hand side of Fig. 6.12, the exclusion
mass bound is not a contour but a point, which is where the theoretical cross section
and the upper limit cross. The observed and 50%-quantile expected upper limits
agree well, statistically confirming the conclusion that the observation is consistent
with a null hypothesis, which was drawn visually in Sect. 6.1.

In all figures, only the experimental uncertainty is used to calculate the 68%- and
95%- quantile expected upper limits. The uncertainty on the theoretical cross section
is displayed for the TChiWg model as a band around the curve for the calculated
cross section, while for the GMSB and T5Wg models, it is implied by the bands
around the observed exclusion contours. In other words, the uncertainty bands on
the observed exclusion curves correspond to where the observed upper limit meets
the calculated cross section if the latter is shifted by $\pm 1\sigma$.

In the GMSB model, the electroweak production cross section is a function of
virtually only the wino mass. Therefore, when the wino is light, a certain number
of signal events are expected regardless of the gluino mass. Conversely, the non-
observation of an excess of events in the signal region results in the bend of the
exclusion curve that stays at a constant value of m_{wino}, as seen in Fig. 6.11. This
result of $m_{\text{wino}} \gtrsim 370\,\text{GeV}$ can be directly compared to the ATLAS result of
$m_{\text{wino}} > 221\,\text{GeV}$ in [9] mentioned in Sect. 2.12.2, marking a clear extension of
the lower bound for the mass of a wino NLSP. Similarly, the mass lower bound
for the gluino of $m_{\text{gluino}} > 815\,\text{GeV}$ at $m_{\text{wino}} = 655\,\text{GeV}$ and $m_{\text{gluino}} > 900\,\text{GeV}$
at $m_{\text{wino}} = 450\,\text{GeV}$ is to be compared with $m_{\text{gluino}} > 619\,\text{GeV}$ in [9] and
$m_{\text{gluino}} > 775\,\text{GeV}$ set by the CMS single photon search [10]. Note that the
convention is to use the observed exclusion where the signal cross section is scaled
down by 1σ when quoting numerical values. In the current case of the GMSB model,
this "conservative" exclusion corresponds to the black dotted curve that is closer to
the origin.

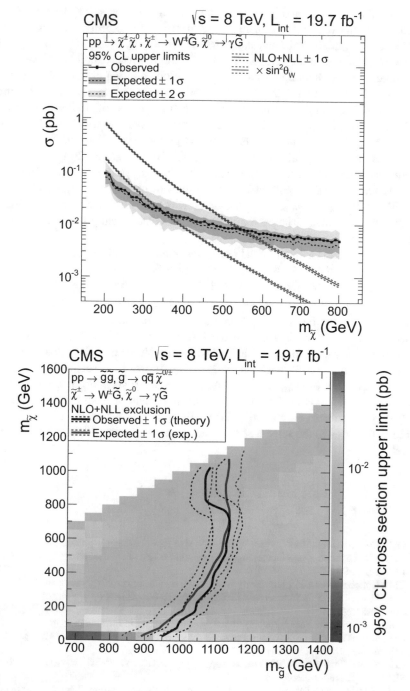

Fig. 6.12 95% CL$_s$ cross section upper limits and theoretical cross section for the TChiWg model (*top*) and the T5Wg model with exclusion contours overlaid (*bottom*)

While the observed and expected exclusion curves generally overlap, a small deviation of the observed exclusion contour from the expectation exists around $550\,\text{GeV} \lesssim m_{\text{wino}} \lesssim 850\,\text{GeV}$ and $870\,\text{GeV} \lesssim m_{\text{gluino}} \lesssim 950\,\text{GeV}$ in Fig. 6.11. This feature is due to a discrepancy between the observed number of events and the background prediction in a single signal region counting bin of $E_T^{\gamma} > 80\,\text{GeV}$, $E_T^{\text{miss}} > 300\,\text{GeV}$, and $100\,\text{GeV} < H_T < 400\,\text{GeV}$ in the $\mu\gamma$ channel. This high-E_T^{γ}, high-E_T^{miss}, mid-H_T bin has an expected background of 0.79 ± 0.10 events, while three events are observed. Since the major contributors to the H_T sum in this model are the (anti)quark jets from the gluino decay $\widetilde{g} \to q\overline{q}\widetilde{\chi}^{\pm}_1(\widetilde{\chi}^0_1)$, for each mass point, a typical H_T value is roughly given by $H_T \sim 2(m_{\text{gluino}} - m_{\text{wino}})$, where the factor of 2 accounts for the two gluinos. Thus the excess in the H_T range $[100, 400]\,\text{GeV}$ mainly affects the region in the $m_{\text{wino}} - m_{\text{gluino}}$ plane with $50\,\text{GeV} < (m_{\text{gluino}} - m_{\text{wino}}) < 200\,\text{GeV}$, which is indeed where the deviation of the contours occurs. The excess is nevertheless statistically insignificant and is considered a statistical fluctuation.

When broken down to individual processes of electroweak and strong production, as simulated in the TChiWg and T5Wg models, $m_{\widetilde{\chi}} < 540\,\text{GeV}$ and $m_{\widetilde{g}} < 1080\,\text{GeV}$ ($m_{\widetilde{\chi}} = 600\,\text{GeV}$) are excluded at 95% confidence level with a 100% branching fraction for $\widetilde{\chi}^0 \to \gamma\widetilde{G}$ and 50% branching fraction for each of $\widetilde{g} \to q\overline{q}\widetilde{\chi}^{\pm}$ and $\widetilde{g} \to q\overline{q}\widetilde{\chi}^0$. For the TChiWg model, a curve for the theoretical cross section scaled down by $\sin^2\theta_W$ is added to the left-hand plot of Fig. 6.12 as a reference to compare this model to more realistic wino-like NLSP models. In particular, the observed limit of $m_{\widetilde{\chi}} > 340\,\text{GeV}$ under this scaled-down cross section is comparable to the quoted limit of $m_{\text{wino}} > 370\,\text{GeV}$ for the GMSB scenario mentioned above. The peculiar deviation of the observed exclusion contour from the expectation in the T5Wg model seen around $m_{\widetilde{g}} - 200\,\text{GeV} \lesssim m_{\widetilde{\chi}} \lesssim m_{\widetilde{g}}$ on the right-hand side of Fig. 6.12 is due to the same upward fluctuation as discussed above for the GMSB model.

References

1. CMS Collaboration: Measurement of the inclusive top-quark pair + photon production cross section in the muon + jets channel in pp collisions at 8 TeV. Technical report CMS-PAS-TOP-13-011 (2014)
2. CMS Collaboration: CMS luminosity based on pixel cluster counting - summer 2013 update. CMS Physics Analysis Summary CMS-PAS-LUM-13-001. CERN (2013). http://cdsweb.cern.ch/record/1598864
3. CMS Collaboration: Search for top-squark pair production in the single- lepton final state in pp collisions at $\sqrt{s} = 8$ TeV. Eur. Phys. J. C **73**(12), 2677 (2013). doi:10.1140/epjc/s10052-013-2677-2. arXiv:1308.1586 [hep-ex]
4. LHC SUSY Cross Section Working Group. https://twiki.cern.ch/twiki/bin/view/LHCPhysics/SUSYCrossSections. Retrieved 1 Jan 2015
5. ATLAS and CMS Collaborations: The LHC Higgs combination group. Procedure for the LHC Higgs boson search combination in summer 2011. Technical report CMS-NOTE-2011-005. ATL-PHYS-PUB-2011-11. CERN, Geneva (2011)

6. Junk, T.: Confidence level computation for combining searches with small statistics. Nucl. Instrum. Methods A 434.CARLETON-OPAL-PHYS-99-01, CERN-EP-99-041 (1999), pp. 435–443. doi:10.1016/S0168-9002(99)00498-2. arXiv: hep-ex/9902006 [hep-ex]

7. Read, A.L.: Presentation of search results: the CL_s technique. J. Phys. G **28**, 2693–2704 (2002). doi:10.1088/0954-3899/28/10/313

8. Moneta, L., et al.: The RooStats project. In: 13th International Workshop on Advanced Computing and Analysis Techniques in Physics Research (ACAT2010). PoS(ACAT2010)057. SISSA, 2010. eprint: 1009.1003 (physics.data-an). http://pos.sissa.it/archive/conferences/093/057/ACAT2010_057

9. ATLAS Collaboration: Search for supersymmetry in events with at least one photon, one lepton, and large missing transverse momentum in proton–proton collision at a center-of-mass energy of 7 TeV with the ATLAS detector. ATLAS Conference Note 2012-144. CERN (2012)

10. CMS Collaboration: Search for supersymmetry in events with one photon, jets and missing transverse energy at $\sqrt{s} = 8$ TeV. Technical report CMS-PAS-SUS-14-004 (2014)

Chapter 7
Conclusion

A search for anomalous production of events with a photon, lepton, and large E_T^{miss} using $19.7\,\mathrm{fb}^{-1}$ of pp collisions at $\sqrt{s} = 8\,\mathrm{TeV}$ recorded in 2012 with the CMS detector at the CERN LHC has been presented. Signal candidate events with large E_T^{miss} and M_T are counted in multiple bins of E_T^{γ}, H_T, and E_T^{miss}. The amount of SM background present in the data is estimated using data-driven methods as well as MC simulation. No excess of events above the expected SM background is observed. The result is interpreted in the context of supersymmetric models with gauge-mediated supersymmetry breaking as limits on the cross section of the pair-production processes of gaugino-like particles that decay subsequently to a W boson and a photon along with nearly massless gravitinos. Assuming that the next-to-lightest supersymmetric particles are the charged and neutral winos (wino co-NLSP scenario), wino and gluino masses below 370 and 820 GeV, respectively, are excluded at 95% confidence level. Compared to the corresponding previous best limits of 221 and 619 GeV [1], this result sets a significantly more stringent constraint on such a scenario of supersymmetry.

Aside from narrowing the parameter space of supersymmetric models, this search establishes background estimation techniques that are generically applicable to many data analyses. In particular, a functional reweighting of control samples to predict arbitrary kinematic distributions of backgrounds from object mis-identification is employed extensively and successfully. Additionally, a template-fit method is utilized to estimate a yet-unmeasured cross section of the main SM background process of $W\gamma$ production, resulting in a reduced systematic uncertainty on the background prediction.

A possible future improvement of this search should involve a more careful evaluation of the contributions from the multiboson and top quark production processes. Particularly, the production of $t\bar{t}$ pair associated with a high-energy

© Springer International Publishing AG 2017
Y. Iiyama, *Search for Supersymmetry in pp Collisions at $\sqrt{s} = 8$ TeV with a Photon, Lepton, and Missing Transverse Energy*, Springer Theses,
DOI 10.1007/978-3-319-58661-8_7

photon will become even more relevant in pp collisions at higher center-of-mass energies, possibly exceeding the contribution from the $W\gamma$ process mentioned above. Such an evaluation may rely on higher-order cross section calculations that are becoming available or some novel data-driven technique.

The non-observation of an excess over the SM prediction in this search does not exclude the wino co-NLSP scenario mentioned above, let alone the supersymmetric standard model with gauge-mediated supersymmetry breaking. A supersymmetric particle with a mass just heavier than what has been excluded in this thesis might be waiting to be discovered. Furthermore, the possibility of a discovery is not only in higher masses. As discussed in Sect. 2.11, the framework of general gauge-mediation admits a large variety of final states, some of which have not been experimentally studied yet. Even for wino co-NLSP scenarios, for instance, the case for long-lived winos resulting in the signature of displaced leptons, photons, and jets has not been addressed thoroughly. The LHC Run II at $\sqrt{s} = 13$ TeV, scheduled to start in the summer of 2015, is an exciting new chapter in the search for supersymmetry at the LHC. A broad set of search programs will probe the newly opened mass range from every possible dimension.

Reference

1. ATLAS Collaboration: Search for supersymmetry in events with at least one photon, one lepton, and large missing transverse momentum in proton–proton collision at a center-of-mass energy of 7 TeV with the ATLAS detector. ATLAS Conference Note 2012-144. CERN (2012)

Printed in the United States
By Bookmasters